THE SCIENCE OF

Stretching

THE SCIENCE OF
Stretching

ALEX REID

THE CROWOOD PRESS

First published in 2017 by
The Crowood Press Ltd
Ramsbury, Marlborough
Wiltshire SN8 2HR

www.crowood.com

British Library Cataloguing-in-Publication Data
A catalogue record for this book is available from the British Library.

ISBN 978 1 78500 260 1

Dedication
For Isla and Machrie who inspire me, make me laugh continuously and amaze me
every day!

Acknowledgements
Thank you to the following contributors within this book. Your efforts and the time
given to help bring things together are greatly appreciated: Peter Court: Peter Court
Media Services (photography); Helen Curzon: Corefit Pilates, Wimbledon Village
(photographic location) and Female Model; Michael Collins: Male model; Emma Britton:
Female Model; Paul Nevin, MSc, BSc, UEFA Pro Licence: Foreword; James Earle,
MSc, BSc (Hons), ASCC: Skeletal Muscle Structure and the Mechanisms of Stretching:
The Science; Chris Bodman MBPsS: Psychological Factors and Routine.

Alex would also like to thank her family and friends who have encouraged and supported her
during the writing and creation of the book.

Typeset by Servis Filmsetting Ltd, Stockport, Cheshire

Printed and bound in India by Replika Press Pvt Ltd

CONTENTS

FOREWORD

Paul Nevin, MSc Performance Coaching, BSc Communications, UEFA Pro Licence, Academy Manager's Licence, First Team Coach, Brighton and Hove Albion.

I was fortunate enough to work alongside Alex for a number of years whilst we were at Fulham Football Club around fifteen years ago. Since those days, my career in professional football has taken me around the world and offered many experiences. As a coach who has worked at youth development and senior level, both in the English Premier League and abroad, the beliefs and practices with regards to stretching seem to be an ongoing enigma to the majority of coaches and players.

I have been exposed to numerous different stretching regimes and concepts at different teams and in different environments, varying from static stretching to dynamic stretching, and from ballistic stretching to no stretching at all. With all of these different stretching modalities and philosophies, which are presented as a seemingly logical argument for inclusion and often with data to support a particular approach for the coaching and support staff to consider, the question is, which one should we choose?

Without exploring the research principals behind each different stretching modality, and perhaps considering coach or player preference, the choice for physical preparation may be slightly ambiguous, haphazard and limited because of current knowledge, hard-headedness, the concept of 'that's what we did' and beliefs. Having trust in the sports science and physical preparation team who prescribe the conditioning to the players is important. It is also important to consider that we are dealing with individual athletes at different points in their careers, with different competition needs, different ages and injury histories, as well as differing environments. For example, I am sure that hot or cold climates and extreme temperatures may also play a part in what is necessary for appropriate physical preparation before training or match play and also for recovery and regeneration.

Although football is a team game, the sport itself is increasingly understanding and respecting the need to treat each team member as an individual with regard to their physical and technical development programmes. As science and research progress, and more and more knowledge and answers are sought and found, I believe that an open and possibly fluid approach to stretching in preparing for competition, preventing injury and ensuring optimal recovery from matches and training is important. The desire

Choosing the appropriate stretching modality has been demonstrated to affect performance variables.

to stretch in order to improve or to maintain flexibility will continue to be practised in its many guises, but knowing when to do it and the correct modality of choice in order to benefit optimally are important, especially in high-performance sport.

In addition to all the stretching scientific theory, current research, practices and specific applications, which you will find in this book, when it comes to most sports and certainly in professional football, it is important to recognize that players will just be comfortable in doing and repeating what feels good for them, especially if their routine brought them a win and three valuable points in the previous game!

INTRODUCTION: WHY DO WE STRETCH?

I have been involved in sport since I can remember, from running around at primary school with my friends, to playing sport at an elite level, to coaching and training athletes in recent years. Throughout this time, the same questions have arisen: Why do we stretch? What type of stretch should I complete? For how long should I hold the stretch? What are the benefits or am I just wasting my time? There seems to be much confusion and ambiguity regarding the rationale and application of different stretching modalities and even whether we should or should not spend any time on stretching at all.

This book will take on a research-based approach, justifying the rationale as to when, why and if we should stretch, and, if we should stretch, what we can expect to be the outcome. The content will address the physiological principals of stretching and any physiological adaptations that occur as a result of stretching the muscle fibres.

I personally believe that people stretch for a number of reasons, such as it's 'what you do' before and after exercise, or it's expected preparation by your coach and teammates. But have you considered the different modes of stretching and why one may be more appropriate than another depending on what your outcome targets may be? Why, for instance, would you select dynamic flexibility rather than static stretching before exercise? Does the muscle temperature or body temperature affect the outcome? Is there a psychological benefit of stretching, such as mental preparation before competition or assisting with prematch nerves? Will we pull a muscle if we don't stretch or pull a muscle if we do?

The Science of Stretching will investigate the reasons for stretching and the rationale for which type of stretch to complete. It will aim to guide you towards a positive outcome with your performance and level of function as a result of effective stretching and mobility exercises.

Mobility is defined as 'the ability to move or be moved freely and easily'.[1] This is in essence what the purpose of stretching may be. From the same resource, to stretch is defined as '(of something soft or elastic) be made or be capable of being made longer or wider without tearing or breaking'. Flexibility is defined as 'the quality of bending easily without breaking', whereas being extensible allows you to accommodate change. These are all ways to describe a form of movement that prepares the human body for performance and function, and ensures an optimal, effective and safe outcome.

Why do we stretch?

In a performance or during physical demands, mobility and extensibility may be more appropriate descriptions when referring to the human body. We want to move freely and easily, especially when performing physical exertions, like dance and sports, and we want to accommodate change, especially when the movements may challenge the full range of motion that the body can achieve, for example, in gymnastics or in a lunge tackle in football.

In elite sport, effective recovery is essential for day to day high performance, but when I look at athletes stretching it appears to be more of a post-session debrief or chat about the weekend, rather than an important component of their conditioning, recovery and fitness. Does this matter? As practitioners, our role surely has to be to educate the athletes and coaches in best practice. The effectiveness and attention to detail that an individual chooses to adopt with this component of fitness may be important and may affect the outcome of their actions. Or is it irrelevant to the outcome? Does which type of stretch you select affect your performance? Are you just holding your quadriceps stretch with poor form and chatting, or are you actually stretching the muscle fibres, tilting your pelvis to increase the range of motion or movement (ROM) and stretch tolerance, and therefore focusing on a positive physical outcome of that specific stretch?

In addition to its application in a sporting environment, stretching can have an important role to play post-operatively and is also prescribed to help with increased mobility after immobilization of a limb. For example, reaching or stretching exercises may be prescribed after a mastectomy to help with circulation and mobility. An understanding of how to manage and increase mobility is an important part of recovery in this situation. Into old age and for sedentary populations, what modality of stretching should be completed for maximum benefit? And how often should these exercises be done? The answers, outcome and rationale will be explored here.

CHAPTER 1

ANATOMY AND THE PHYSIOLOGICAL PRINCIPLES OF MOVEMENT

Stretching the muscle unit effectively involves increasing its length and tension. In order to achieve this effectively, we need to understand the workings of the muscles and levers of the skeleton. The first thing we need to understand is the structure of muscle fibres.

Muscle Fibres and their Role within the Body

There are around 700 muscles in the skeletal system; each is an organ – constructed of skeletal muscle tissue, blood vessels, tendons and nerves – that allows us to generate force. Muscle tissue has four main properties:

- excitability: the ability to respond to stimuli
- contractibility: the ability to contract
- extensibility: the ability to be stretched without tearing
- elasticity: the ability to return to its normal shape.

All four of the above properties are important when it comes to stretching muscles. We need to create excitability and to ensure that the muscle reacts to the stimuli of movement. The levers of the skeletal system work with co-contraction of the agonist and antagonist muscles, so when we stretch the quadriceps, for example, we require the hamstrings to contract or shorten. We need good extensibility so that the muscle can stretch without tearing and we would like it to return to its pre-stretch form and not become deformed as a result of the stretch. In some instances, we would like to increase muscle length and ROM if possible, and we will address this later on.

Based on certain structural and functional characteristics, muscle tissue is classified into three types: skeletal, cardiac and smooth:

- Skeletal muscle (or striated muscle) is responsible for locomotion and general movement. Skeletal muscle tissue can be made to contract or relax by conscious control (voluntary).
- Cardiac muscle (heart) – the contraction is completed without thinking about the muscular action and is therefore involuntary.
- Smooth muscle is also an involuntary muscle. The muscles line the walls of the arteries to control blood pressure, control the digestion of food by causing movement of the intestine and the urinary bladder, for example.

Types of Muscle

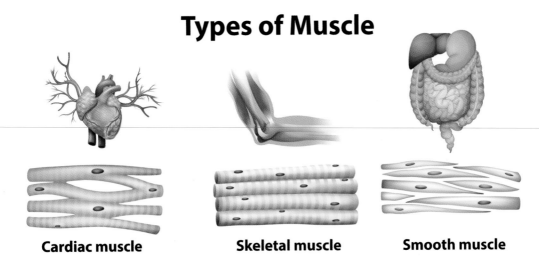

Cardiac muscle   **Skeletal muscle**   **Smooth muscle**

Types of muscle tissue: cardiac, skeletal and smooth.

Three Main Types of Muscle Fibre			
Fibre Type:	**Type I Fibres**	**Type IIA Fibres**	**Type IIB Fibres**
Contraction time	Slow	Fast	Very fast
Size of motor neuron	Small	Large	Very large
Resistance to fatigue	High	Intermediate	Low
Activity used for	Aerobic	Long-term anaerobic	Short-term anaerobic
Force production	Low	High	Very high
Mitochondrial density	High	High	Low
Capillary density	High	Intermediate	Low
Oxidative capacity	High	High	Low
Glycolytic capacity	Low	High	High
Major storage fuel	Triglycerides	CP, glycogen	CP, glycogen

BrianMac Sports Coach[3]

We also need to consider fascia, as this is an important structure within the skeletal system. Fascia is the soft tissue component of the connective tissue system. It interpenetrates and surrounds muscles, bones, organs, nerves, blood vessels and other structures. Fascia is an uninterrupted, three-dimensional web of tissue that extends from head to toe, from front to back, from interior to exterior.[2]

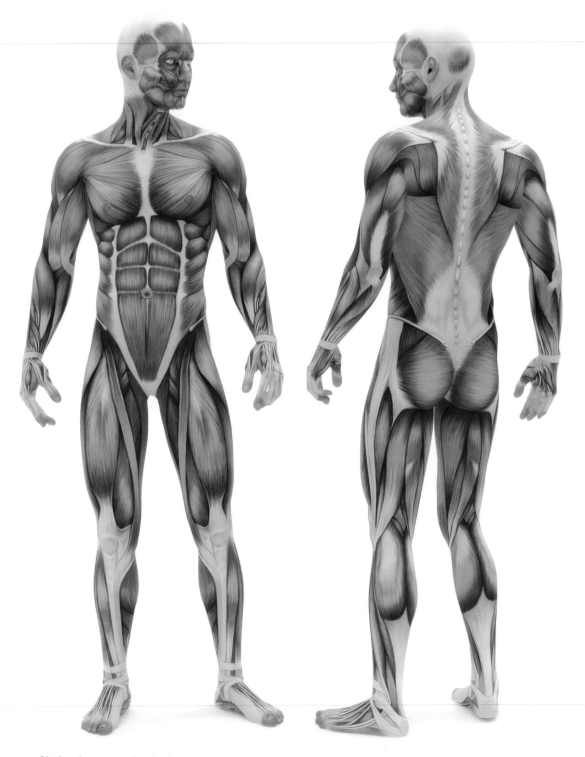

Skeletal system: the depicted white areas are fascia within the skeletal system.

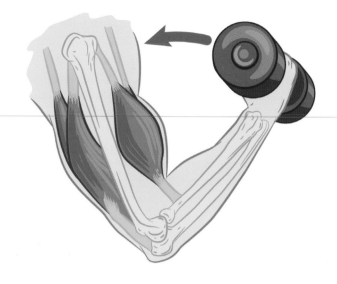

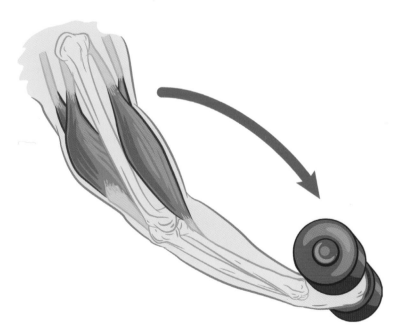

A biceps curl is an example of a lever in the human body.

Fascia is responsible for maintaining the structural integrity of the skeleton and for providing support and protection. It also acts as a shock absorber.

Skeletal muscles contain thousands of muscle cells, or muscle fibres, which run between tendons. They have a capacity to contract and extend, which allows for movement. There are three types of muscle fibre and each one has a specific role within muscular function, as shown in the table on page 11: Type I, Type IIA and Type IIB.

Most skeletal muscles within the body are a mixture of all three types of muscle fibres,

13

but their proportion varies depending upon the action of the muscle. For example, postural muscles of the neck, back and leg have a higher proportion of type I fibres. This allows these muscles to remain active as they have a high resistance to fatigue and are aerobic in nature, so can function at a constant activity level. Muscles of the shoulders and arms are not constantly active but are used intermittently, usually for short periods, to produce large amounts of tension such as in lifting and throwing. These muscles have a higher proportion of type I and type IIB fibres.

Even though most skeletal muscle is a mixture of all three types, all the skeletal muscle fibres of any one motor unit are the same. In addition, the different skeletal muscle fibres in a muscle may be used in various ways, depending upon the need. For example, if only a weak contraction is needed to perform a task, only type I fibres are activated by their motor units. If a stronger contraction is needed, the motor units of type IIA fibres are activated. If a maximal contraction is required, motor units of type IIB fibres are activated as well. Activation of various motor units is determined in the brain and in the spinal cord. Although the number of the different skeletal muscle fibres does not change, the characteristics of those present can be altered by training and load.

Movement within the skeletal system is achieved by the use of levers. There are three types of lever within the human body: first-class lever; second-class lever; and third-class lever. They are defined by the relative position of the three elements of the lever: the effort (E); the position of the fulcrum (F); and the load, or resistance (R). The muscles attach to the skeleton via tendons and as a muscle shortens through a contraction, it will cause movement along the lever, or a contraction of the muscle. It is necessary to have a fulcrum to work with the lever to allow for movement.

In the human body, the joints act as fulcrums and the bones as levers. As one muscle shortens, the opposite muscle will lengthen with concentric or eccentric movement. This is the basic concept of a lever within the human body, using agonist and antagonist muscles. It is important to consider the angle of pull so as to understand what may incite a stretch response in the particular part of the body you are aiming to stretch. If the goal is to elongate a muscle from its insertion, then by considering the angle of pull within the lever you can achieve the desired effect of stretch within the chosen muscle group. Anteriorly tilting your pelvis to elicit an increased stretch in your hip flexors would be an example whereby our body awareness can help in manipulating the stretch sensation. Indeed, knowing how the body moves and how this relates to flexibility and function is important when prescribing any exercise, including stretching.

Muscular physiology is an area where it would be best to turn to an expert to explain the concept as simply as possible. James Earle, a Sports Scientist at St Mary's University, Twickenham, has kindly put together the relevant information to enhance our understanding of skeletal muscle structure (see the Feature Box). He also touches on the mechanisms of stretching.

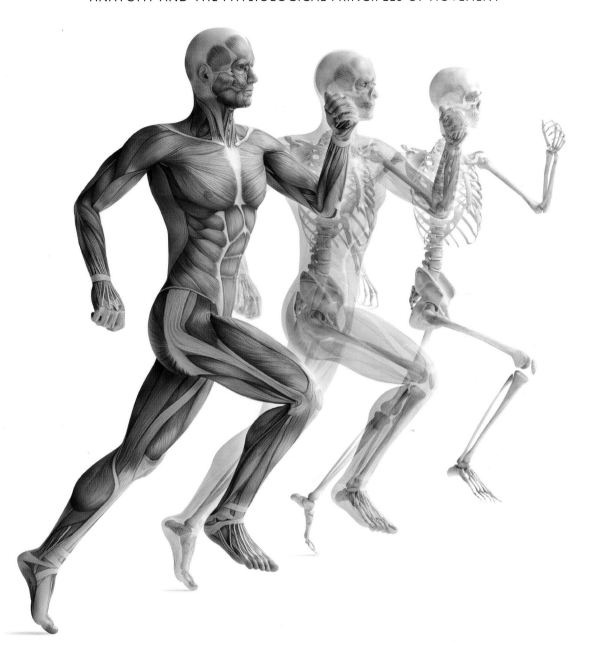

Knowing how the body moves helps with effective exercise prescription.

'SKELETAL MUSCLE STRUCTURE AND THE MECHANISMS OF STRETCHING: THE SCIENCE' BY JAMES EARLE, MSC, BSC (HONS), ASCC

The human body has over 660 skeletal muscles; each is a complex structure of muscle fibres surrounded by connective tissue. A muscle fibre consists of contractile proteins called myofibrils, which are constructed in a linear series of sarcomeres. A sarcomere is the smallest functional unit of a muscle that can contract and is composed of proteins that interact together. One of these proteins, actin, forms thin myofilaments, whereas the other protein, myosin, forms thick myofilaments. The structural differences between actin and myosin give sarcomeres a 'banded' appearance.

A thin layer of connective tissue termed the endomysium surrounds each individual muscle fibre. The perimysium surrounds a group of muscle fibres and this is known as a fascicle. Together, bunches of fascicles form a muscle, which is surrounded by a dense fascia of connective tissue called the epimysium. This protective sheath transitions at the ends of a muscle into tendons and connects it to bone.

As a muscle contracts, the muscle fibres shorten as the myofilaments slide towards the centre of each sarcomere; this is known as the sliding filament theory. Each thin filament contains hundreds of active sites ready to bind to myosin contained in the thick filament. These sites are covered by a troponin-tropomyosin complex, which prevents exposure when a muscle is at rest. When calcium is released into the muscle via an action potential, it alters the troponin-tropomyosin complex and exposes the active sites. Hydrolysis of ATP (adenosine triphosphate) provides energy to move the myosin molecules on the thick filaments, which can then attach to the thin filaments and interact, forming a cross-bridge. This formation creates a pulling action, termed a power stroke, and shortens the sarcomere.

Once the sarcomere has shortened, there is no active mechanism for relaxation. Instead, a muscle fibre returns to its resting length through a combination of processes:

■ Elastic forces – energy spent contracting/stretching the muscle and tendons is recovered as they rebound to

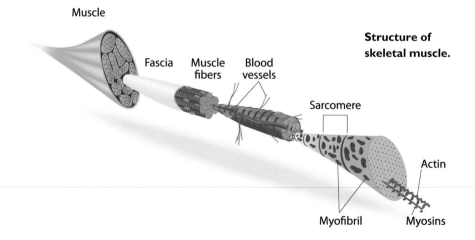

Muscle

Structure of skeletal muscle.

Fascia · Muscle fibers · Blood vessels

Sarcomere

Actin

Myofibril · Myosins

their original states; a similar response occurs during a stretch shortening cycle.

■ Antagonist muscle contraction – the contraction of an opposite muscle can restore a muscle to its resting length.

■ Gravity – this acts as a supporting mechanism that assists the antagonist muscle's contraction.

A resting muscle is stretched into a position by tendons that already maximize its force production. This length–tension relationship of a muscle describes the optimal length for force generation. Therefore, changing the length of a muscle will consequently alter its contractile force.

During muscle contractions there are protective sensors that provide sensory feedback to the central nervous system (CNS) on limb position, joint angle and muscle tension. These are called proprioceptors. Muscle spindles located throughout the muscle detect changes in muscle length and tension beyond normal range and initiate a stretch reflex to prevent muscle fibre damage. Golgi tendon organs located at the muscle–tendon junction sense excessive tension during a muscle contraction and send signals to the spinal cord to reduce force output, termed 'reflex inhibition'.

Stretching is used to improve joint ROM. The muscular adaptations to stretching are mechanical, followed by neural changes, the opposite to strength training. However, the extent to which these changes affect flexibility is unclear, with a number of theories proposed. Skeletal muscle behaves with both elastic and viscous properties, that is, when a muscle is lengthened it will always return to its resting length and the tensile force is rate and force dependent. When a stretch is applied, the muscle's inherent elasticity is altered as the muscle relaxes; this is called viscoelastic stress relaxation.[a] This reduction in the muscle's resistance allows it slowly to 'creep' in length. This response is evident during a repeated effort of proprioceptive

neuromuscular facilitation (PNF) stretching, when you can move the muscle past its initial stretch point. Studies[b,c,d] have found an increase in fascicle length following stretching, supporting a change in the viscoelastic properties of a muscle. However, as stiffness or resistance of a muscle is a continually changing property, the effects are temporary, for example 30min to 2hr.[e,f,g] This has led to the theory being questioned regarding having a significant effect on understanding flexibility.

Another possible mechanical adaptation is that stretching with a contractile element could increase the number of sarcomeres in series. Similarly, eccentric strength training (where the muscle is trained in a lengthened position) has been shown to alter the length–tension relationship,[h,i] also suggesting an increase in sarcomeres in series. This has clinical implications in individuals recovering from injury who have had the muscle immobilized in a shortened position.

Other mechanical theories include deformation of the connective tissue and neuromuscular relaxation. However, both have been questioned, as any disruption to connective tissue would modify the length–tension relationship, which is not conclusive in the research. Furthermore, stretch reflexes are only activated during a very rapid and short stimulus and are, therefore, unlikely to cause neuromuscular relaxation in a muscle during stretching.[a]

Rather than a structural change to a muscle, stretching may simply cause an alteration in the perception of the stretch and improve stretch tolerance.[a,c,j,k,l,m] This has been shown with individuals having their stretch tolerance improved within six weeks of stretching.[j,k,l]

PNF stretching with an isometric contraction, or a single leg deadlift where the hamstrings perform an eccentric contraction are examples of lengthening a muscle with a contractile element.

To summarize, we have explained skeletal muscle anatomy and the physiological mechanisms of stretching. The research is

(continued)

Single leg straight dead lift: eccentric strength training that can help with increased ROM.

PNF stretch: isometric contraction to increase ROM.

still unclear as to the magnitude of effect these mechanisms have on flexibility, with different theories proposed. What is known is that stretching likely causes both mechanical and neural adaptations that contribute to flexibility and joint ROM. Whilst the differences in stretching methods have not been discussed here, active and passive stretching have demonstrated comparable results to increases in stretch tolerance, while showing different effects on neuromuscular performance[n] and reductions in muscle and tendon stiffness.[o] This hints towards a greater adaptive response from active stretching. In addition, strength training[p] and specifically eccentric training have been shown to be effective methods for increasing flexibility.[q] Therefore, a comprehensive individualized training programme, incorporating a variety of techniques, will develop and maintain flexibility.

To summarize:

- A sarcomere is the smallest contractile element of a muscle and contains the myofilaments actin and myosin; their interaction can explain muscle contraction by the sliding filament theory.
- A muscle is protected from excessive force, tension and length by proprioceptors and muscle spindles. A tendon is protected from excessive tension by Golgi tendon organs.
- Stretching may increase the length of a muscle temporarily by altering its viscoelastic properties and/or improving the tolerance to the stretch.

References

a Weppler, C. and Magnussan, P. (2010), 'Increasing Muscle Extensibility: A Matter of Increasing Length or Modifying Sensation?' *Physical Therapy*, 90, pp 438–49.

b Abellaneda, S., Guissard, N. and Duchateau, J. (2009), 'The Relative Lengthening of the Myotendinous Structures in the Medial Gastrocnemius During Passive Stretching Differs Among Individuals', *Journal of Applied Physiology*, 106, pp 169–77.

c Blazevich, A., Cannavan, D., Waugh, C., Miller, S., Thorland, J., Aagaard, P. and Kay, A. (2014), 'Range of Motion, Neuromechanical and Architectural Adaptations to Plantar Flexor Stretch Training in Humans', *Journal of Applied Physiology*, 117, pp 452–62.

d Theis, N., Korff, T., Kairon, H. and Mohagheghi, A. (2013), 'Does Acute Passive Stretching Increase Muscle Length in Children with Cerebral Palsy', *Clinical Biomechanics*, 28, pp 1,061–7.

e Martini, F., and Nath, J. (2009), *Anatomy and Physiology*, 8th Edition, Chapter 10 (San Francisco: Pearson Education Inc.).

f Mizuno, T., Matsumoto, M. and Umemura, Y. (2013), 'Decrements in Stiffness are Restored within 10 Min', *International Journal of Sports Medicine*, 34, pp 484–90.

g Ryan, E., Beck, T., Herda, T., Hull, H., Hartman, M., Costa, P., Defreitas, J., Stout, J. and Cramer, J. (2008), 'The Time Course of Musculotendinous Stiffness Responses Following Different Durations of Passive Stretching', *Journal of Orthopaedic and Sports Physical Therapy*, 38, pp 632–9.

h Aquino, C., Fonseca, S., Gonçalves, G., Silva, P., Ocarino, J. and Mancini, M. (2010), 'Stretching versus Strength Training in Lengthened Position in Subjects with Tight Hamstring Muscles: A Randomized Controlled Trial', *Manual Therapy*, 15, pp 26–31.

i Brughelli, M. and Cronin, J. (2007), 'Altering the Length–Tension Relationship with Eccentric Exercise', *Sports Medicine*, 37, pp 807–26.

j Folpp, H., Deall, S., Harvey, L. and Gwinn, T. (2006), 'Can Apparent Increase in Muscle Extensibility with Regular Stretch be Explained by Changes in Tolerance to Stretch?' *Australian Journal of Physiotherapy*, 52, pp 45–50.

k Ben, M., and Harvey, L. (2010), 'Regular Stretch Does Not Increase Muscle Extensibility: A Randomized Controlled Trial', *Scandinavian Journal of Medicine and Science in Sports*, 20, pp 136–44.

l Konrad, A. and Tilp, M. (2014), I'ncreased Range of Motion After Static Stretching is not Due to Changes in Muscle and Tendon Structures', *Clinical Biomechanics*, 29, pp 636–42.

m Magnusson, S., Simonsen, E., Aagaard, P., Dyhre-Poulsen, P., McHugh, M. and Kjaer, M. (1996), 'Mechanical and Physical Responses to Stretching with and without Preisometric Contraction in Human Skeletal Muscle', *Archives of Physical Medicine and Rehabilitation*, 77, pp 373–78.

n Minshull, C., Eston, R., Bailey, A., Rees, D., Gleeson, N. (2014), 'The Differential Effects of PNF versus Passive Stretch Conditioning on Neuromuscular Performance', *European Journal of Sport Science*, 14, pp 233–41.

o Kay, A., Husbands-Beasley, J. and Blazevich, A. (2015), 'Effects of PNF, Static Stretch and Isometric Contractions on Muscle–Tendon Mechanics', *Medicine and Science in Sport and Exercise*. Ahead of print.

p Morton, S., Whitehead, J., Brinkert, R. and Caine, D. (2011), 'Resistance Training vs. Static Stretching: Effects on Flexibility and Strength', *Journal of Strength and Conditioning Research*, 25, pp 3,391–8.

q O'Sullivan, K., McAuliffe, S., and DeBurca, N. (2012), 'The Effects of Eccentric Training on Lower Limb Flexibility: A Systematic Review', *British Journal of Sports Medicine*, 46, pp 838–45.

CHAPTER 2

STRETCHING A MUSCLE: DO WE KNOW WHAT IS HAPPENING?

A number of theories have been proposed within the literature to explain what happens when a muscle is stretched and how an increase in muscle extensibility and flexibility is observed after stretching. Many of the theories and research papers advocate that a mechanical increase in length of the stretched muscle is what happens as a result of the stretch. More recently, however, a theory has been proposed suggesting instead that increases in muscle extensibility and ROM are due to a modification of sensation only, indicating that although there are measurable mechanical changes within the tissue, these improvements occur because the individual is able to lengthen their muscles further because of decreased discomfort at the end range and have an increased stretch tolerance.

Weppler and Magnussen[4] discuss the mechanical variations that may affect stretch as well as the sensation concept. They discuss the mechanical theories, which include viscoelastic deformation, plastic deformation, increased sarcomeres in series and neuromuscular relaxation. Their rationale is that because muscle comprises deformable material, its length measurement at a given moment in time is always dependent upon the amount of tensile force applied.

The mechanical variations that may affect stretch and extensibility of a muscle include the following.

VISCOELASTIC DEFORMATION

Skeletal muscles are considered to be viscoelastic. Increases in muscle extensibility observed immediately after stretching are due to a lasting viscoelastic deformation. Like solid materials, they demonstrate elasticity by resuming their original length once tensile force is removed. Yet, like liquids, they also behave viscously because their response to tensile force is rate and time dependent.[5,6]

Özkaya et al.[5] explain how an immediate increase in muscle length can occur due to the viscous behaviour of muscles whenever they undergo stretch of a sufficient magnitude and duration or frequency. This increased length is a viscoelastic deformation because its magnitude and duration are limited by a muscle's inherent elasticity. With static stretching, when the muscle is held in position for a period of time, its resistance to stretch diminishes over time and declines; this is called viscoelastic stress relaxation. Özkaya et al. discuss the constant load of stretching, such as stretching that uses a fixed torque, and discuss how this can be used to evaluate the property of creep. Creep occurs when mechanical length gradually increases in response to a constant stretching force.

However, there is another school of thought that viscoelastic deformation is not a reason for lasting increases in muscle length and extensibility, and that the magnitude and duration of the length increases vary depending upon the duration of the stretch and the type of stretching applied. With the application and prescription of stretching typical of that practised in rehabilitation and sports, the biomechanical effect of viscoelastic deformation can be quite minimal and so short lived that it may have no influence on subsequent stretches according to Weppler and Magnusson.[7]

PLASTIC DEFORMATION OF CONNECTIVE TISSUE

Another theory suggests that increases in human muscle extensibility observed immediately after stretching are due to plastic or permanent deformation of connective tissue. This mechanical theory is discussed by Chan et al.[8] and Draper et al.,[9] among others.

The classical model of plastic deformation requires a stretch intensity sufficient to pull connective tissue within the muscle past the elastic limit and into the plastic region of the torque/angle curve, so that once the stretching force is removed, the muscle would not return to its original length, but would remain permanently in a lengthened state. Beyond the plastic state is rupture and then permanent deformation would occur. Plastic deformation requires very heavy stretch loads, which is perhaps a limiting factor in an applied setting.

INCREASED SARCOMERES IN SERIES

Weppler and Magnusson[10] discuss the animal studies relating to increased sarcomeres in series as a result of stretching and immobilization of limbs. They cite and discuss how muscles adapt to a new functional length by changing the number and length of sarcomeres in series in order to optimize force production at the new functional length.

The research on animals overall has been generalized to consider whether human skeletal muscle would have the same response to shortening or lengthening. Tabary et al.[11] looked at plaster casts on cats and demonstrated that the number of sarcomeres in series of a muscle can be changed by prolonged immobilization in extreme positions. That is, when muscles are immobilized in fully extended positions, there is an increase in the number of sarcomeres in series and if muscles are shortened, then there is a decrease in sarcomeres in series and a decrease in muscle length. We know that this occurs when a limb is immobilized with a fracture, for example, and how difficult it may be to regain full ROM after the cast has been removed. It is still vague as to whether there is an histological change in human skeletal muscle with the sarcomeres in series and this is certainly an area of potential research.

NEUROMUSCULAR RELAXATION

Neuromuscular relaxation is an involuntary contraction of muscles due to a neuromuscular stretch reflex, which may lead to elongation of tissue. In order to increase muscle extensibility, it has often been proposed that slowly applied static stretch, used alone or in combination with therapeutic techniques associated with proprioceptive neuromuscular facilitation (PNF), stimulates neuromuscular reflexes that induce relaxation of muscles undergoing static stretch and allows an elongation of the muscle beyond the previous end range. The research is varied on this being a reason for increased muscle length and we will discuss PNF stretching later in the book.

SENSORY THEORY

Many of the studies that Weppler and Magnusson[12] discuss consider the sensory theory, with stretching and end-point pain or discomfort being a limiting factor.

Generally, the studies use end-point pain as a marker for range of motion and after a stretching protocol the subjects feel more comfortable increasing the ROM because they are familiar with the sensation. These studies therefore suggest that increases in muscle extensibility observed immediately after stretching and after short-term stretching programmes (three to eight weeks) are due to an alteration of sensation only and not to an increase in muscle length.

Halbertsma and Göeken[13] discuss the stretch tolerance from a group of subjects with short hamstrings, demonstrating that after a four-week stretching programme with one group plus a control group, the subjects with short hamstrings showed a slight but significant increase in the extensibility of

Hurdler demonstrating good functional range of movement.

the hamstrings, accompanied by a significant increase of the stretching moment tolerated by the passive hamstring muscles. However, the elasticity of the hamstrings remained the same. Halbertsma and Göeken therefore concluded that stretching exercises do not make short hamstrings any longer or less stiff, but only influence the stretch tolerance.

An individual's sensation is an important component when considering what is happening during a stretch and may provide information regarding how the muscle is routinely used during functional activity. If the muscle is pushed beyond its normal limits and ROM occasionally, then the sensation of stretch will be higher and more intense than if a certain or extreme ROM is completed on a regular basis. Form follows function with the human body, so the body adapts and perhaps it is sensation and increased comfort, or perhaps it is a functional change in mechanical properties within the muscle fibres that allows for an increased stretch or ROM. It is certainly an area that can still be debated.

A hurdler in athletics or a professional footballer would be conditioned to perform within extreme ranges of functional movement that would be unaccustomed movement patterns for others. Therefore when these athletes perform explosive hurdle jumps in track hurdles or make a lunge tackle in football, it would be within their normal functional range and be uneventful, but a similar movement would most likely cause injury in someone unconditioned for this type of movement pattern.

Footballer demonstrating the range of movement demands in a tackle.

CHAPTER 3

ASSESSMENT OF FLEXIBILITY AND MOBILITY

What is classified as optimal muscle length? When an athlete gets assessed by their physiotherapist, strength and conditioning coach or sports medicine team, what constitutes tight hamstrings? What is normal? Or hypermobile? What indicates an increased injury risk? Do we compromise strength if our muscle length is too much?

If we are going to address and consider our levels of mobility and flexibility and the capacity of stretch within the body, we need to consider relevant assessments to measure our current range of motion within a joint to help identify any deficiency and these assessments need to be objective, valid, reliable and repeatable.

If we have a deficiency left to right, do we have an increased injury risk? Does this need to be addressed via improved flexibility, mobility or increased ROM? Or increased strength? What is a normal range of motion? Or normal variance left to right? If it isn't broke, then why should we fix it? Or should we fix everything that we find?

In the sporting and medical world, there are a number of screening assessments that may be used with a team of athletes or individuals. These assessments will provide an objective measure to the clinician as to whether the individual is within 'normal' parameters and,

if not, a corrective exercise programme may be prescribed.

Below is a selection of screening assessments and tools, plus the rationale for selection depending upon the goals and targets of the individual or practitioner. Some assess isolated ROM of a limb or joint, while others assess more functional capacity around a joint. Each has its pros and cons and may or may not be relevant, depending upon the objectives and goals of assessment.

FUNCTIONAL MOVEMENT SCREEN

The original Functional Movement Screen (FMS) was launched in 1995 and is the screening tool used to identify limitations or asymmetries in seven fundamental movement patterns. The seven movement patterns have been identified as key functional movement qualities in individuals with no current pain complaint or known musculoskeletal injury. The FMS is to be used for healthy, active people and for healthy and inactive people who want to increase physical activity.

The seven screens for the Functional Movement Screen are as follows:

1. Deep Squat
2. Hurdle Step-over
3. In-line Lunge
4. Shoulder Mobility
5. Active Straight Leg Raise
6. Trunk Stability Push-up
7. Rotary Stability.

The above screens can be conducted by a health care professional, strength and conditioning coach, physical education teacher and so on. The tests are assessed very simply on a 0–3 point scale and the screen is relatively easy to do once the assessor is familiar with the movement patterns. Corrective exercises are then prescribed to address any imbalances.

In recent years, Functional Movement Systems, who created the original FMS, has launched the Selective Functional Movement Assessment (SFMA). This movement-based diagnostic system is designed for the clinical assessment of seven fundamental movement patterns in clients or athletes with known musculoskeletal pain. The assessment provides an efficient method for systematically finding the cause of symptoms, not just the source, by logically breaking down dysfunctional patterns and diagnosing their root cause as either a mobility problem or a stability/motor-control problem.

The screen is aimed to be conducted by professional qualified clinicians (chartered physiotherapists, osteopaths, orthopaedic medical practitioners and so on) to help them to assess their clients' musculoskeletal evaluation and assist in further corrective exercises and diagnosis.

The seven SFMA are as follows:

1. Cervical Spine Movement Assessment
2. Upper Extremity Movement Pattern of the Shoulder
3. Multi-segmental Flexion Assessment
4. Multi-segmental Extension Assessment
5. Multi-segmental Rotation Assessment
6. Single Leg Stance Assessment
7. Overhead Deep Squat Assessment.

There are sub-assessments on most of the above screens as well to look at movement through range and these assessments should be completed by a qualified practitioner.

The common denominator with both of the screening systems above is that mobility and global whole-body movement patterns are assessed. This means that a lack of ROM or reduced flexibility may hinder a movement pattern on the screen, scoring a lower measure, or may cause pain or impingement. This will be highlighted from the screen and corrective exercises prescribed by the trained clinician. You can find out more at www.functionalmovement.com.

The scientific data on the FMS has been analysed by Kraus et al.[14] with the conclusion that results suggest that the FMS is a reliable screen if the assessor is educated and has solid experience in the screening process and movement patterns. Studies that Kraus et al. reviewed clearly illustrate the FMS's limited ability to predict athletic performance. However, to predict injury risk in team sports, the FMS total score is supported by moderate scientific evidence so may be a worthwhile addition as a screening tool, especially in a team environment where corrective exercise can be prescribed and regularly monitored.

SIT AND REACH TEST

The sit and reach test is a common measure of flexibility. It specifically measures the flexibility of the lower back and hamstring muscles.

This test involves sitting on the floor with

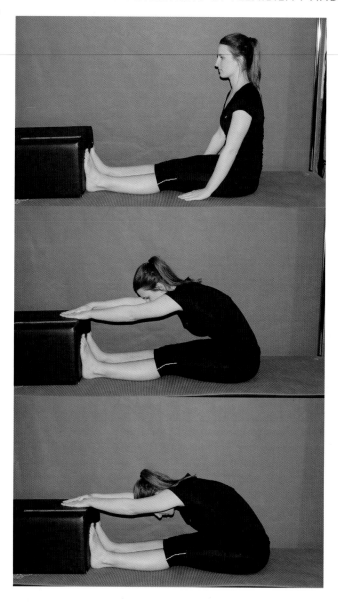

Sit and reach test.

the legs out straight ahead. Bare feet are placed with the soles flat against the box, shoulder-width apart. Both knees are held flat against the floor by the tester, if required. With hands on top of each other and palms facing down, the subject reaches forward along the measuring line as far as possible. After three practice reaches, the fourth reach is held for at least 2sec while the distance reached is recorded.

This test is important because tightness in this area is implicated in lumbar lordosis, forward pelvic tilt and lower back pain. This test was first described by Wells and Dillon.[15] The test has been used in many settings, including in police forces and fire departments and for military personnel, in addition to schools and other communities and health-related fitness environments.

There are a number of modified sit and

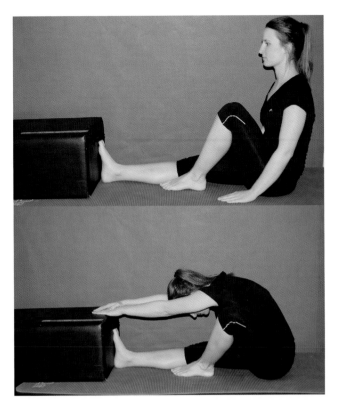

Backsaver sit and reach test.

reach tests that try to standardize the measure and address any potential inaccuracy. With this test, for example, people with long arms and/or short legs would get a better result, while those with short arms and/or long legs are at a disadvantage. The test therefore needs to be valid and reliable for all shapes and sizes! Holt et al.[16] present a number of modified protocols for stand and reach/sit and reach, in a quest to ensure that there are limited factors which may affect the outcome of the test.

Holt et al. found that a number of modified sit and reach protocols allow the indirect assessment of the influence of the four major muscle groups which affect sit and reach scores, these being erector spinae, hip rotators, hamstrings and gastrocnemii. Laboratory evaluations of kinesiology students and human cadavers have confirmed the influence of other muscle groups with the test also (Burke[17]).

Holt et al. also noted that if the position of the appropriate muscle groups (hamstrings, gastrocnemii and external rotators in this case) is slackened or changed, the relative contribution to the achieved sit and reach measures by each isolated muscle group can be identified. This opens the discussion as to whether muscles should be tested in isolation or as a group working together.

There have been other modifications, for example the Backsaver sit and reach test developed by FitnessGram, which is designed to assess the fitness levels of children in grades K–12 in the USA. FitnessGram assesses three general components of health-related physical fitness, which have been identified as important to overall health and function. The Backsaver sit and reach test is different to the original sit and reach test. The main difference from the original test is that the athlete bends at one knee and then performs the sliding movement for measurement.

In other variations of the test. the main variable that changes is the measurement markings on the sit and reach box itself. Using where the toes meet the box as zero makes sense, but this is where the long arms and short legs conundrum comes into effect. So there is the modified sit and reach test, which addresses this issue, as the zero mark is adjusted for each individual, based on their sitting reach level.

If you do use this test to assess flexibility, make sure that you use the correct normative data to compare because of all the variations that are possible. And be sure to retest with the same protocol!

STRAIGHT LEG RAISE TEST

The straight leg raise test is very similar to the active knee extension test below and is commonly used by physiotherapists and sports medicine practitioners to assess the range of hip flexion of each leg with the knee fully extended and assessing the sciatic nerve, lower lumbar and upper sacral nerve roots for irritation.

To complete the test, the patient is positioned in a supine position (on their back) on a clinician's couch. The clinician lifts the patient's symptomatic leg by the posterior ankle while keeping the knee in a fully extended position. The clinician continues to lift the patient's leg by flexing at the hip until pain is elicited or end range is reached. The measure of range is either observed and noted, or measured with a goniometer. This test is an indicator of neurological pain, assessing the sciatic nerve, lower lumbar and upper sacral nerve roots for irritation. The

test is positive if significant back pain, or pain in the lower extremity, is present. This test is more of a clinical measure than a performance measure, but it is important for athletes or individuals with a history of lumbar spine issues.

ACTIVE KNEE EXTENSION TEST

(See photos on page 30) The purpose of this test is to assess the range of active knee extension and hamstring ROM in a position of hip flexion, as required in running and kicking.

The subject lies supine, head back and arms across the chest. The hip is passively flexed until the thigh is vertical. The knee may be flexed and relaxed at this point. The individual needs to then maintain this thigh position throughout the test, with the opposite leg in a fully extended position on the couch. The foot of the leg being tested is kept relaxed, while the knee is actively extended and straightened until the point when the thigh begins to move from the vertical position, indicating that a full ROM of hamstring extension has been reached.

The measurement, recorded in degrees, is the minimum angle of knee flexion with the thigh in the vertical position. If the leg is able to be fully straightened, the angle would be recorded as 0 degrees. Any degree of flexion will be recorded as a positive number, for example 10, 20 degrees and so on. In cases where the full knee extension is achieved without thigh movement, the knee is flexed and the thigh is moved to 30 degrees past the vertical position, and the knee again straightened and re-evaluated. The angle of knee flexion at which the thigh begins to move is again recorded. This would indicate good hamstring flexibility.

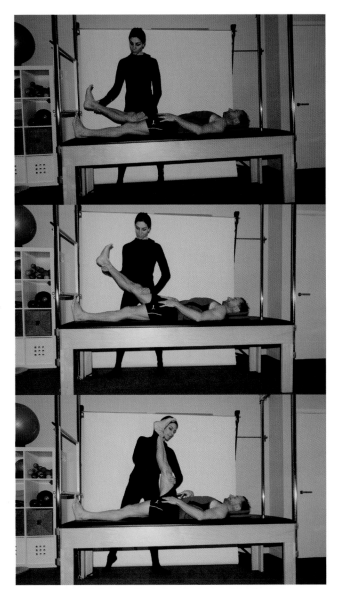

Straight leg raise.

THOMAS TEST

The Thomas test, Iliacus test or Iliopsoas test, as it is also known, is used to measure the flexibility of the iliopsoas muscle group, the rectus femoris, as well as the tensor fascia latae and the sartorius; (see photos on pages 31 and 32).

The patient is asked to lie supine on the treatment couch. The examiner checks for lordosis in the spine, which is a predictor of a tight hip flexor. The examiner then flexes one hip, bringing the knee to the chest of the patient and asks the patient to hold or hug the knee tight so as to help stabilize the pelvis and flatten out the lumbar region. If the hanging leg, which is the one being tested (the leg on the table) does not have a hip flexion contraction, it will remain on the testing table. If it raises up off the table, this means that a contracture is present. This is measured by

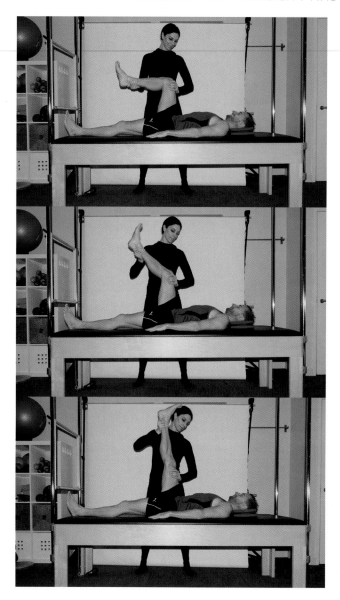

Active knee extension test.

a goniometer, if present, so as to provide a ROM to compare each limb.

The test can also be performed with the starting position for the patient where both knees would be fully flexed to the chest on the treatment couch and the patient then slowly lowers the leg being tested to see if the leg makes it to the treatment couch. This is the Modified Thomas test. Lack of full hip extension with knee flexion less than 45 degrees indicates iliopsoas tightness. If full extension is reached in this position, it would indicate rectus femoris tightness. If any hip external rotation is observed, it may indicate ITB tightness. The test can therefore provide some important objective measures for anterior thigh tightness.

The test reviews the range of motion of the hip based upon the aforementioned muscle's ROM and mobility. Reduced ROM or bilateral

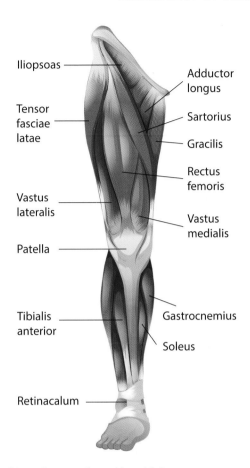

Iliopsoas

Tensor fasciae latae

Vastus lateralis

Patella

Tibialis anterior

Retinacalum

Adductor longus

Sartorius

Gracilis

Rectus femoris

Vastus medialis

Gastrocnemius

Soleus

Musculature of anterior thigh.

asymmetry may indicate that there may be some muscle imbalance, or indication for intervention, or a requirement for improved ROM or stability.

Kim[18] examined the test–retest reliability of the Modified Thomas test using lumbopelvic stabilization. In their study, the authors used thirteen subjects who all had hip flexor tightness in order to look at the reliability of three different ways to measure the Modified Thomas test. The participants underwent the Modified Thomas test under three conditions: first, the general Modified Thomas test was performed; second, active lumbopelvic stabilization by internal fixation was accomplished using a biofeedback pressure of 40mmHg during the Modified Thomas test; and third, passive lumbo-pelvic stabilization by external fixation was accomplished with the examiner's hand on the right side of the pelvis (dominant leg) during the Modified Thomas test. The knee-joint angle under the three conditions was determined using the Simi motion-analysis software.

The researchers found that the active lumbo-pelvic stabilization and passive lumbo-pelvic stabilization methods for the Modified Thomas test were more reliable than the general Modified Thomas test method. The score for the active lumbo-pelvic stabilization method was 2.35 degrees, which indicated

Thomas test.

Modified Thomas test.

that a real difference existed between two testing sessions, compared with the scores for the passive lumbo-pelvic stabilization at 3.70 degrees and the general Modified Thomas test method, which measured 4.17 degrees. The research concludes that lumbo-pelvic stabilization is one of the considerations to allow for precise measurement of the Modified Thomas test and may help to minimize measurement error when evaluating hip flexor tightness. Most experienced physiotherapists or sports medicine practitioners will complete the Thomas test with a visual comparison rather than with a goniometer.

Peeler et al.[19] also looked at the clinical reliability of the Modified Thomas test for evaluating the flexibility of the rectus femoris muscle about the knee joint. Three experienced athletic therapists with an average of 12.67 years of sport medicine expertise assessed rectus femoris flexibility using pass/fail and goniometer scoring systems. A retest session was completed seven to ten days later to check reliability. Peeler et al. determined that there were generally poor levels of reliability for pass/fail scoring results and fair to moderate levels of reliability for goniometer data measurement. They call into question the statistical reliability of the Modified Thomas test and provide clinicians with important information regarding its reliability limits when used for the clinical assessment of the flexibility of the rectus femoris muscle about the knee joint in a physically active population. Certainly, further research is needed to ascertain the variables that may confound the statistical reliability of this orthopaedic technique, which is commonly used to assess patients in a clinical setting and to prescribe resultant rehabilitation.

Among many experienced physiotherapists, the Thomas test is widely used to assess a patient's anterior thigh ROM. The practitioners' experience allows for a visual review regarding how the limb falls during the assessment and, in combination with clinical detail, patient history and any additional objective measures or markers, these would all work in combination and lead to any appropriate exercise prescription and analysis of ROM.

WEIGHT-BEARING LUNGE TEST: KNEE TO WALL

The weight-bearing lunge test is a common assessment of flexibility in the calf and also of the ROM in the ankle joint. It is also commonly known as the knee to wall test.

The athlete stands with their hips square

Knee to wall test.

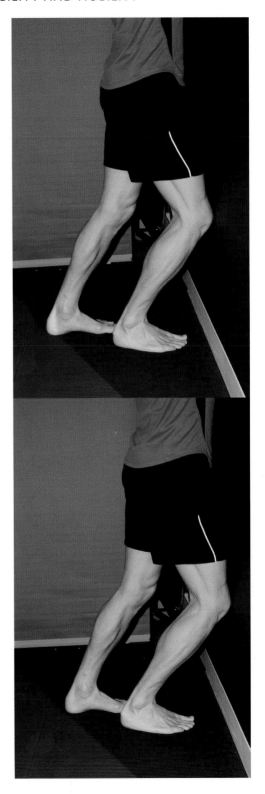

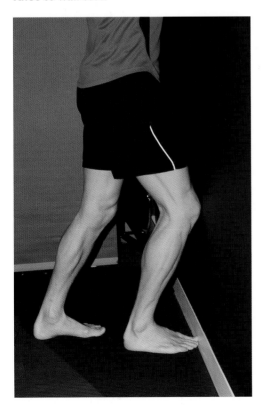

to the wall, positioned with the testing leg in front. The heel stays flat on the floor, the knee touches the wall and a marker or tape is placed to see how far from the wall the athlete can go before the heel comes up off the floor and the knee can no longer touch the wall. Once the knee can no longer touch the wall, or the heel comes up, this indicates the achieved ROM, or the level of ankle dorsiflexion range. The test is completed on both sides for comparison and the score achieved is relative to the individual.

This is an important measure and a marker that chartered physiotherapists, sports medicine practitioners and athletics trainers will often use to asses ankle range of motion as a screening tool, but also after injury when the joint may be thicker and stiff and have a marked reduction in ROM. Prospective studies have shown it to be predictive of injury (Pope et al.[20]; Gabbe et al.[21]) which makes it a valuable objective screening tool.

A significant discrepancy on either side may affect a person's gait and cause problems up the kinetic chain in the body or lead to injury, so this can be an important assessment relating to functional flexibility and mobility, especially if presented with an ankle injury history, calf problems or plantar fascia issues. The test has also been shown to have very good reliability and repeatability (Bennell et al.[22]). However, individual asymmetry should be measured and compared, rather than using normative data as a comparison.

There should always be a reason and good rationale for measures and assessment to be completed with all types of physical assessments and tests. All selected tests need to be objective, valid, reliable and repeatable. But on occasion we may want to have some functional 'stiffness', as some ankle stiffness may be advantageous in an athlete, for example. In which case, we may not want to increase the ROM in 'stiff' ankles with a low knee to wall score if the athlete is asymptomatic and performing well. Ankle or leg stiffness therefore may a desirable trait in sports!

Consider a 100m sprinter: if they land with poor ankle stiffness, they lose their body's force into the ground, as it will be absorbed instead of being fired back into the body to help propel it forward. Ground

CASE STUDY: STIFF ANKLES

One of the athletes that I work with has a very poor knee to wall score and his heels raise off the floor when he squats down, which indicates tight calf muscles. He has very stiff ankles and his knee to wall measures on both sides are very limited and could be classified as poor, with limited ROM, but he is clinically quiet, with no injury issues or any relevant injury history: he is just made that way. He is also the quickest athlete over 40m in his athlete group. If we worked on his calf flexibility and improved his knee to wall score by increasing his calf muscle length, would he become slower as a result of having less 'stiffness' in his ankles?

These are the kind of questions that we need to consider when screening, testing and making decisions. It is very important when we discuss the application of any screening or assessment result with our athletes. What are we going to do with our objective measures? We may prescribe exercises or stretches to overcome any deficiency, but is this appropriate? If it isn't broke, should we feel like we should fix it?

contact time will be long and slow. If there is an element of appropriate ankle stiffness, upon landing the force will be transferred directly back into the muscles and used to create more force. This is called the stretch-shortening cycle and James Earle touched on this in his article.

The stretch-shortening cycle is an active stretch (eccentric contraction) of a muscle, followed by an immediate shortening (concentric contraction) of that same muscle. It allows the body to react in a fast, reactive way, so is very desirable in sports performance. We are looking for short ground contact time and explosivity with our elite athletes, and even with endurance athletes who need a 'kick' at the end of a race, so we need to consider how best to harness this for our athletes with what we ask them to do in their conditioning and training.

So when we screen our athletes, we need to consider the function and demands of their sports, their injury history and any other relevant test results. We need to ensure that we achieve a balance of strength and appropriate mobility or flexibility without affecting an athlete's optimal physicality for enhanced performance.

NEURAL TESTS

Slump Test

Innervation of the body is essential for daily life. If the nervous system is compromised in

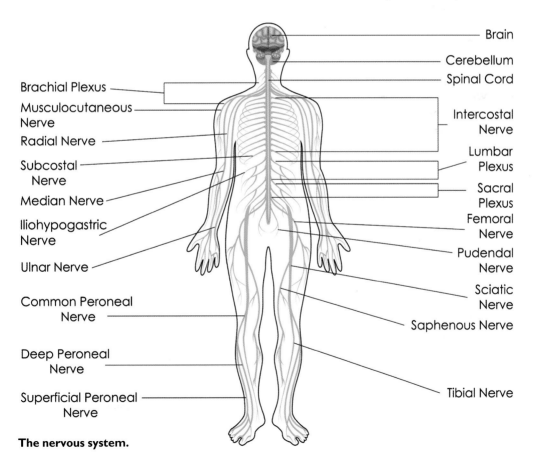

Brain
Cerebellum
Spinal Cord
Brachial Plexus
Musculocutaneous Nerve
Radial Nerve
Subcostal Nerve
Median Nerve
Iliohypogastric Nerve
Ulnar Nerve
Common Peroneal Nerve
Deep Peroneal Nerve
Superficial Peroneal Nerve
Intercostal Nerve
Lumbar Plexus
Sacral Plexus
Femoral Nerve
Pudendal Nerve
Sciatic Nerve
Saphenous Nerve
Tibial Nerve

The nervous system.

any way there will be a 'short circuit' within the body, which may lead to pain, dysfunction or abnormal sensation. The nerves extend across the joints around the body and they need to slide, glide, bend, elongate and withstand compression as the body moves in daily activity along their pathway. On occasion, nerves can become more alert and sensitive to the stress placed upon them, for example in the presence of inflammation or injury. If this is the case, just normal movement can cause pain, numbness, tingling or other signs of nerve distress. This is called adverse neural tension and may cause injury as the body contracts the muscles around the nerves to protect them.

THE NERVOUS SYSTEM

The slump test is used to assess adverse neural tension in patients with low back and hamstring injuries. It involves tensioning the neural tissues without additional hamstring stretch. This is achieved by flexing the cervical and thoracic spine during hamstring stretch along the posterior chain.

The patient is seated upright with their hands held together behind their back. The practitioner instructs the patient to flex their spine with bad posture, which causes the 'slump' position. This is followed by neck flexion, when they will place their chin to their chest. The practitioner can then place their hand on top of the patient's head and ask the patient to perform a knee extension, whilst dorsiflexing the foot on one side. Once the patient has completed a number of repetitions, they are instructed to return the neck to a neutral position. The test is considered positive if symptoms are increased in the slumped and flexed position, and decreased as the patient moves out of neck flexion.

The results of the test may indicate if an individual is experiencing symptoms related to nerves adhering to various tissues while travelling throughout the body. They may be experiencing increased sensation of stretching, pain or other neurological sensations within the area of adhesions. So, some people may feel the neural tension in the glutes, hamstrings, calf muscles or the lumbar spine, depending upon where the adhesions and restrictions are. Poor ROM on one side or the other may also be identified when comparing bilaterally.

Another use for the test is detecting lumbar disc herniation. With the flexed lumbar spine and hip completed simultaneously with the extended leg extension position, the sciatic nerve and its respective nerve roots are put under tension and will indicate the potential diagnosis of a disc herniation. The results of the test should be interpreted based upon the patient's pain/symptoms and clinical presentation. It is especially important to assess an individual's neural tension with the slump test if they complain of hamstring pain, which could be misdiagnosed as a muscular strain, which will be discussed later in the book.

Slump Knee Bend Test and Prone Knee Bend Test

The slump knee bend test and the prone knee bend test both assesses the glide of the femoral nerve anteriorly and have been used to help indicate radicular pain, or pain originating as a result of irritation to the spinal structures and may indicate irritation of the nerve roots at L2–L3 due to the innervation of the femoral nerve. In addition, pain in the anterior thigh may indicate a tight/strained quadriceps muscle or neural tension of the femoral nerve.

The test can be conducted prone (with the patient on their front) as the 'prone knee bend test', or also with the subject on their

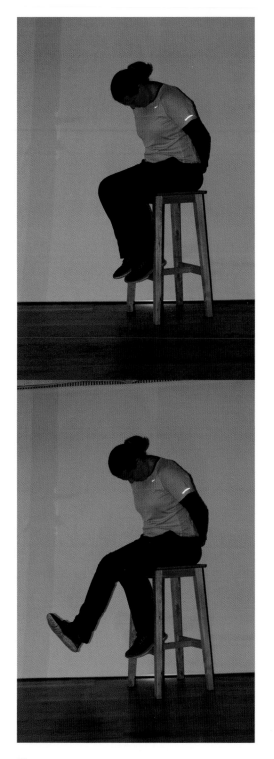

Slump test.

Prone knee bend test.

side. If the subject is on their side, it is also possible to illicit spinal flexion, which may lead to more accurate results in neural tension assessment, much the same as in the seated slump test.

For the prone knee bend test the practitioner passively flexes the subject's knee to end range and maintains it there for 45sec. The hip should not be rotated. Symptoms and pain are considered throughout the test.

However, the slump knee bend test is thought to have superiority over the prone knee bend test in differentiation between symptoms arising from neural versus non-neural tissues because of the addition of the spinal flexion component.

The slump knee bend test is performed as the subject is side lying, hugging the underside leg (but not fully flexing it), with their cervical and thoracic spines flexed. The practitioner stands behind the subject supporting the upper leg to maintain a neutral hip position. The subject's upper knee should be flexed and their hip extended to the point of evoked symptoms. Once symptoms are evoked, the subject is asked to extend their neck and the practitioner monitors changes in symptoms and resistance to hip movement before ending the test. The results should be compared to the non-symptomatic side and clinical detail in conjunction with the test outcome is once again very important.

The position of knee flexion puts a stretch on the femoral nerve and its nerve rootlets due to the nerve passing on the anterior side of the lower extremity. Should the femoral nerve become adherent to the tissues it passes by in the lower extremity, pain or other neural symptoms may be produced in that area. A positive test is when the spinal flexion is returned to neutral and the pain/sensation subsides.

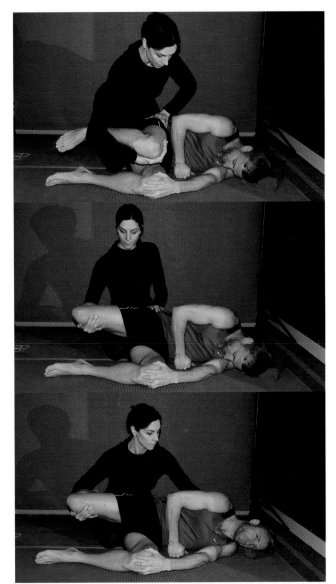

Slump knee bend test.

BEIGHTON SCORE AND BRIGHTON CRITERIA

If someone is hypermobile and has problems with regular dislocation or joint pain, the Beighton Score may help with diagnosis.

The Beighton modification of the Carter and Wilkinson scoring system has been used for many years as an indicator of widespread hypermobility by practitioners. It consists of a series of five tests, the results of which can add up to a total of nine points. An individual's score is established as follows:

- one point if the subject can place their palms on the ground while standing with legs straight

- one point for each elbow that bends backwards
- one point for each knee that bends backwards
- one point for each thumb that touches the forearm when bent backwards
- one point for each little finger that bends backwards beyond 90 degrees.

If the Beighton Score is four or more, it is likely that the subject has joint hypermobility.

The Brighton Criteria takes into account the Beighton Score, but also considers other symptoms, such as joint pain and dislocated joints, and how long these have been in existence. There are major and minor Brighton Criteria which may be relevant to an affected individual. More information on the Beighton Score and hypermobility can be found at www.hypermobility.org. Often strength training may be prescribed for hypermobile individuals to shorten a muscle and help reduce hypermobility around the joint, which may be beneficial.

FLEXITEST

De Araújo[23] developed a method for the measurement and evaluation of flexibility in a range of subjects, old and young, to create normative data and called it the Flexitest. This is a method for the evaluation of twenty joint movements. The normative data and work that de Araújo completed was originally published is 2005, presenting normative data for the joints measured, while taking into account age and gender differences. In this time, the researchers had evaluated and created normative data on over 4,700 non-athlete subjects.

The author is from Brazil and the Flexitest is used in the Brazilian Air Force as part of its fitness assessment, which implies credibility and accuracy. The Flexitest is also referenced in a number of articles as a measurement technique for assessing flexibility, mainly in older subjects. The normative data created is for male and female subjects ranged between five and ninety-one years.

Originally introduced in 1980 and with redesigned evaluation maps published in 1986, the Flexitest consists of the assessment of mobility with the use of a scale from 0 to 4. By adding the individual results of the twenty joint movements assessed, it is possible to obtain a global score called the Flexindex.

The method consists of measuring and evaluating the maximum passive ROM of the twenty body joint movements. This includes the ankles, knees, hips, trunk, wrists, elbows and shoulders. Each of the movements is measured in a growing discontinuous scale from 0 to 4, thus comprising a total of five possible values. Evaluation maps relating to age and gender have been created and the value is taken when the maximum passive ROM is achieved on each joint. The central trend of measurement is 2, then 1 and 3 values are less frequent, while the extreme values, 0 and 4, are very uncommon.

The total score from all twenty body joint movements is obtained to provide a global index of flexibility, or Flexindex. Statistically, Flexindex results for men and women differ from ten years of age. In practical terms, women tend to be more flexible than men from five years of age, with a difference of approximately 5 per cent; this difference is intensified after puberty, leading to between a 10 to 15 per cent difference and even more in the third decade of life, with around 20 per cent gender difference. After sixty years of age, the differences tend to be even greater in women, reaching between 20 to 40 per cent more flexible than the value obtained for the Flexindex in men. This normative data is no doubt useful in a clinical setting where normative data may be used in comparison for certain populations. *Flexitest: An Innovative Flexibility Assessment Method* by C.G. de Araújo[24] provides further reading.

There are a number of other screening protocols to assess flexibility and mobility. Some are clinical measures for orthopaedic professionals, while other are performance-related measures. The key is that, as with any screening process, you need to ensure that each test is valid, reliable, objective and repeatable. You may use a goniometer to measure ROM, or a passive feel, as with the Flexitest, or a more objective measure, as in the sit and reach test when the outcome is an actual measure. However you measure flexibility and ROM, the assessment has to provide relevant information, which leads to the answers you are seeking. It has to be repeatable so that you can measure, re-measure and observe any changes in ROM or score achieved, so that you can assess how successful your corrective exercise plan has been, or otherwise.

CHAPTER 4

DIFFERENT STRETCHING MODALITIES AND TECHNIQUES

PASSIVE AND ACTIVE STRETCHING

Passive stretching means that you are using some sort of outside assistance to help you achieve a stretch. This assistance could be your body weight, a strap, leverage, gravity, another person or a stretching device. Commonly combined with static stretching, for example within a standing quadriceps stretch,

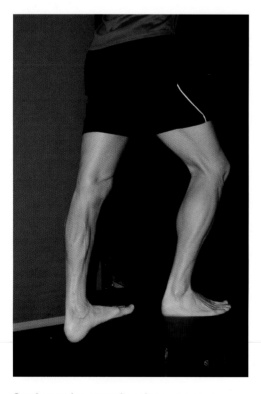

Static-passive stretch: calves on a step.

Hamstring stretch on a bench.

when you hold your ankle to increase the ROM in your anterior thigh, this would be classed as a passive-static stretch. Most stretches have an external force placed upon them, via a body part, or a bench, step or the floor for a calf stretch, for example.

Active stretching is when you contract the opposing muscles to stretch the targeted muscle with no outside force. For example, if you were to sit with your legs extended and dorsiflex your ankles, this would stretch the calves by tightening the muscles on the top of the shins, the anterior tibialis muscle. All stretches are either passive or active. Passive and active stretches can both be either dynamic or static.

The definition of static-passive stretching and static-active stretching and dynamic-active stretching and dynamic-passive stretching seem interchangeable when the research is

Dynamic-passive stretch: up and down calf stretch on step.

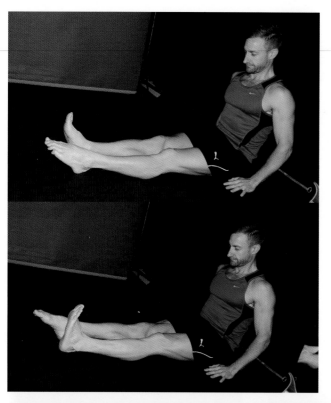

Static-active stretch: calves via ankle dorsiflexion.

Seated hamstring stretch.

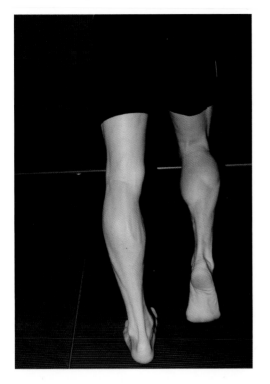

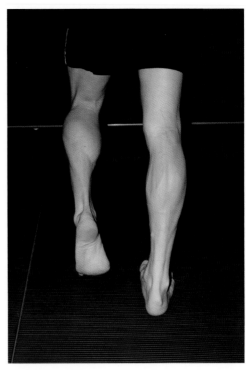

Dynamic-active stretch: calves up and down via ankle dorsiflexion.

Dynamic hamstring stretch.

Dynamic hamstring stretch (cont).

reviewed. Both static and dynamic stretching will be addressed and discussed in detail below, addressing research-based principles and the rationale for appropriate application.

DELAYED ONSET MUSCLE SORENESS AND RECOVERY STRATEGIES WITHIN EACH STRETCHING MODALITY

The golden chalice for any sports person, whether recreational, professional or elite, is the ability to train on consecutive days at an appropriate intensity to illicit a positive training outcome in order to enhance performance.

If we are stiff and sore as a result of unaccustomed or high-intensity exercise, which causes delayed onset muscle soreness (DOMS), then our quality of training and capacity for additional load may be compromised, yet we need to overload our body and challenge ourselves every day to become the best. So, how do we achieve this in relation to stretching and can stretching help to alleviate DOMS?

Within each stretching modality subject below is a subsection, if relevant, addressing how that particular stretch may affect DOMS and if it is a beneficial modality.

STATIC STRETCHING

Static stretching involves gradually easing into the stretch position with a group of

Static stretch: gastrocnemius stretch.

muscles and holding the position for a prescribed duration, feeling a gentle stretch while holding. The amount of time for which a static stretch is held may depend upon both a person's objectives and the findings described below. Does the research even indicate that it is of any use at all?

Should static stretching be part of your warm-up or cool-down and, if so, how long should the stretches be held for? Does this or any type of stretching affect the post-exercise micro-tears in the muscles, with the goal being to realign the muscle fibres and help to pump out any waste products resulting from the exercise session and therefore reduce DOMS and aid recovery? Or does it not help?

Many people stretch before or after exercise, with the goal being to reduce the risk of injury and muscle soreness after exercise, or to enhance athletic performance. Herbert et

al.[25] reviewed a total of twelve randomized studies from the Cochrane Bone, Joint and Muscle Trauma Group specialized register, and from this data concluded that stretching before or after exercise, or both before and after exercise, does not reduce DOMS in healthy adults.

Often in static stretching you are advised to move further into the stretch sensation position as the stretch sensation subsides in order to increase your ROM. This may be advisable for developmental flexibility, when the desired outcome is to improve long-term muscle length and ROM. But can you maintain this gain in muscle length? How frequently do you have to stretch to gain any increased length? And the great debate with static stretching is how long should the muscle be held under tension?

The answers tend to depend upon your

goals from the training session or competition. Are you about to complete a maximal sprint or vertical jump assessment and therefore require a high power output and maximal force, or are you about to complete a 10K run or a marathon? Specificity of stretching and applying the correct principles have been shown via research to affect performance either negatively or positively. An understanding of optimal frequency, duration of stretch and the desired performance outcome as a result of this choice is therefore very important.

Siatras et al.[26] investigated the acute effect of different static-stretching durations on quadriceps' isometric and isokinetic peak torque production. The fifty participants were randomly allocated into five equivalent-sized groups and were asked to perform a stretching exercise of different duration. The varied durations were: no stretch; 10sec stretch; 20sec stretch; 30sec stretch; and 60sec stretch. The knee flexion range of motion and the isometric and concentric isokinetic peak torques of the quadriceps were measured before and after a static-stretching exercise.

The research indicated that significant knee joint flexibility increases and significant isometric and isokinetic peak torque reductions have been shown to occur after 30sec and 60sec of quadriceps static stretching. Isometric peak torque was reduced by 8.5 per cent after 30sec and 16.0 per cent after 60sec of static stretching. Isokinetic peak torque after 30sec and 60sec of stretching was reduced by 5.5 per cent and 11.6 per cent at 60 degrees/sec and by 5.8 per cent and 10.0 per cent at 180 degrees/sec. Siatras et al. therefore recommend that static-stretching exercises of a muscle group for more than 30sec of duration be avoided before performances requiring maximal strength.

This reduction in performance is probably due to neuro-mechanical properties within the muscles. Ogura et al.[27] also discuss the duration of static-stretching holds and concluded that short duration (<30sec) of static stretching did not have a negative effect on the muscle force production. Anything greater than 30sec may therefore reduce performance in relation to maximal strength output.

Beckett et al.[28] looked at the effects of static stretching on repeated sprint and change of direction. They examined the effects of static stretching during the recovery periods of field-based team sports on subsequent repeated sprint ability and change of direction speed performance. They concluded that, from their protocol, an acute bout (4min) of static stretching of the lower limbs during recovery periods between efforts may compromise repeated sprint ability performance, but has less effect on change of direction speed performance.

Kokkonen et al.[29] investigated the influence of static-stretching exercises on specific exercise performances. His team looked at a group who completed a regular 10-week, 40min × 3 days each week static-stretching routine designed to stretch all the major muscle groups in the lower extremity. They then compared their outcomes with a control group who did not participate in any regular exercise routine during the study. A number of performance variables were measured in each group before and after the ten-week cycle. The static-stretching group showed significant average improvements in all of the performance measures by purely addressing their static-stretching component of fitness, as stated above.

These performance measures included improvements in the following: flexibility (sit and reach measure; 18.1 per cent); standing long jump (2.3 per cent); vertical jump (6.7 per cent); 20m sprint (1.3 per cent); knee flexion, one repetition maximum (15.3

per cent); knee extension, one repetition maximum (32.4 per cent); knee flexion endurance (30.4 per cent); and knee extension endurance (28.5 per cent). The control group showed no improvement in any performance measures. The conclusion of the authors suggests that chronic static-stretching exercises by themselves can improve specific exercise performances. It is possible that persons who are unable to participate in traditional strength training activities may be able to experience gains through stretching, which would allow them to transition into a more traditional exercise regimen.

This is very interesting information for older populations or less active subjects who may benefit from stretching classes, yoga and Pilates for example, as a health-related exercise benefit. This study demonstrates a long-term benefit on the bodies of the subjects, rather than an immediate acute physical effect that many of the other studies address, so it may therefore indicate long-term benefits that should be considered for long-term athlete development and also for older or more sedentary populations.

Behm et al.[30] looked at the effect of acute static stretching on force, balance, reaction time and movement time. They looked at acute bouts of static stretching for 45sec at the point of discomfort and then assessed maximal voluntary isometric contraction (MVIC) force of the leg extensors, static balance using a computerized wobble board, reaction and movement time of the dominant lower limb, and the ability to match 30 per cent and 50 per cent of MVIC forces with and without visual feedback.

There were no significant differences in the decrease in MVIC between the stretch and control conditions, or in the ability to match submaximal forces. However, there was a significant decrease in balance scores with the stretch (decreasing 9.2 per cent) compared with the control (increasing 17.3 per cent) condition. Similarly, decreases in reaction (5.8 per cent) and movement (5.7 per cent) time with the control condition differed significantly from the stretch-induced increases of 4.0 per cent and 1.9 per cent, respectively. From their research, it appears that an acute bout of stretching impaired the warm-up effect achieved under control conditions with balance and reaction/movement time. This is important information for activities that demand excellent reaction times and balance, such as tennis, gymnastics or dance. It could also affect injury potential if proprioceptive awareness is compromised, as with measures on the electronic wobble board.

However, it should be noted that these are results from a research paper; I have rarely seen a stretch held at a point of discomfort for 45sec in a performance environment. Therefore, on an applied note, this information, although of interest, may not affect what would be prescribed to athletes. But it is interesting to acknowledge that static stretching may lead to these deficiencies in performance and is certainly worth being aware of. On occasion, I have seen individuals hold stretches or some trainers push through to a point of discomfort with stretches with their athletes before exercise and it is important to consider the potential outcome of these actions and the subsequent effect on performance.

Allison et al.[31] looked at running economy and whether static stretching had any effect on running economy parameters: oxygen uptake; minute ventilation; energy expenditure; respiratory exchange ratio; and heart-rate responses during 10min of treadmill running at 70 per cent VO_2. Although the research indicated that there were neuromuscular changes as a result of the stretching protocol, leading to an increase in the sit and reach score and a decrease in isometric strength and counter-movement jump

Dancer demonstrating a large functional range of movement.

height, the physiological parameters that affect running economy were unchanged as a result of static stretching. The protocol that the authors used included a static-stretching intervention where each leg was stretched unilaterally for 40sec with each of eight different exercises and this was repeated three times. Allison et al. concluded that the results suggest that prolonged static stretching does not influence running economy at 70 per cent VO_2 during a 10min treadmill run, despite changes in neuromuscular function.

However, Lowery et al.[32] looked at running economy over a 1-mile hill run at 5 per cent incline on a treadmill to be completed as fast as possible. The demands of running uphill require more power output than flat running, so it is interesting to note that this research

indicates that with this protocol of six lower-body stretches for three 30sec repetitions prior to the run (with the control group completing nothing prior to the run), running economy was negatively affected. The participants for the Lowery et al. study would have completed the run at a significantly higher VO_2 demand than the participants in the study by Allison et al. because of the intensity of the run and the prescribed treadmill incline.

Lowery et al. concluded that the time to complete the run was significantly less (6:51 ± 0:28min) in the non-stretching condition, as compared with the stretching condition (7:04 ± 0:32min). A significant condition by time interaction for muscle activation existed, with no change in the non-stretching condition (pre- 91.3 ± 11.6 mV to post- 92.2 ±

With the demands of the game, should the squash player complete static stretches before his or her match?

12.9 mV), but increased in the stretching condition (pre- 91.0 ± 11.6 mV to post- 105.3 ± 12.9 mV). A significant condition by time interaction for ground contact time (GCT) was also present, with no changes in the non-stretching condition (pre- 211.4 ± 20.8 ms to post- 212.5 ± 21.7 ms), but increased in the stretching trial (pre- 210.7 ± 19.6 ms to post- 237.21 ± 22.4 ms). A significant condition by time interaction for flexibility was found, which was increased in the stretching condition (pre- 33.1 ± 2 to post- 38.8 ± 2), but unchanged in the non-stretching condition (pre- 33.5 ± 2 to post-35.2 ± 2). So, once again, the performance variables have been negatively affected, one can conclude, by the static stretching of 3 × 30sec holds prior to the bout of exercise.

Lowery et al. looked at different variables with their study compared to Allison et al.. Lowery et al. used an actual time-affected outcome and running the prescribed distance as fast as possible, while Allison et al. studied the physiological components at a submaximal level. So if we consider both papers, we can agree that: flexibility is increased in both examples (via the sit and reach test); and that running economy measured via a gas-analysis system at a submaximal level is unaffected, but as the physical demands increase, the negative effect of static stretching before high-intensity exercise bouts are highlighted.

Lowery et al. state in their conclusion that their study findings indicate that static stretching decreases performance in short endurance bouts by as much as 8 per cent, while increasing GCT, which, when we refer to the stretch shortening cycle, is something to avoid if we want to be quick and explosive. Therefore, they conclude that coaches and athletes may be at risk of decreased performance after a static-stretching bout and that static stretching with holds of 30sec should be avoided before a short endurance event.

Wilson et al.[33] also discussed the effects of static stretching on energy cost and running endurance performance. They looked at a combination of a steady run consisting of a 30min 65 per cent VO_2 max preload followed by a 30min performance run, in which participants ran as far as possible without viewing distance or speed. The total duration was 60min. The stretching group completed 16min of static stretching using five exercises for the major lower body muscle groups, whereas the non-stretching group completed 16min of quiet sitting before the run. Performance was measured as distance covered in the performance run and also total calorie expenditure was determined for the 30min preload run.

Performance was greater in the non-stretching group (6.0 +/− 1.1km) versus the stretching group (5.8 +/− 1.0km) condition, with greater energy expenditure during the stretching group compared with the non-stretching group condition (425 calories +/− 50 versus 405 calories +/− 50 kcals). Wilson et al.'s findings suggest that stretching before an endurance event may lower endurance performance, measured here as distance covered, and it may increase the energy cost of running, demanding a higher calorific output.

Research by Yamaguchi et al.[34] has demonstrated that there is a loss in power output from the muscle if a stretch is held for too long prior to exercise (static stretching). Their research used a protocol of 4 × 30sec

holds with six types of static stretches. With running in particular being a dynamic, powerful event, a loss in power output is not desirable; for this reason, athletes should not perform static stretches with holds up to 30sec prior to exercise.

DePino et al.[35] looked at the duration of maintained flexibility gains in knee joint range of motion on same day static hamstring stretching. The purpose of the study was to determine the duration of hamstring flexibility gains as measured by an active knee-extension test after the cessation of an acute stretching protocol. The participants performed 4 × 30sec static stretches, with 15sec rest between each repetition. They were then measured at 1, 3, 6, 9, 15 and 30min after cessation of the stretching protocol. Analysis of the study indicated significant improvement of knee extension range of motion in the experimental group, lasting 3min after cessation of the static-stretching protocol. Subsequent measurements after 3min were not statistically different from the baseline. This is important to consider when selecting a stretching protocol.

This protocol indicates that any change in ROM is short lived and will return to pre-stretching levels 3min after finishing the stretch. Looking at the research by Kokkonen et al.,[36] we know that there may be chronic gains from regular stretching, but that this appears to be a long-term commitment to change and not a quick fix. More research is certainly warranted in this area.

If we refer back to our muscle physiology and what happens to a muscle when it is stretched, we can understand why the muscle may return back to its pre-stretch values. We know that skeletal muscles are considered to be viscoelastic and that increases in muscle extensibility observed immediately after stretching may be due to a lasting viscoelastic deformation. Like solid materials, they demonstrate elasticity by resuming their original length once tensile force is removed. DePino et al. discusses that the reason for a return to pre-stretch values is because of a temporary creep effect, in which the viscoelastic component of the hamstring was not deformed enough to produce a permanent change.

DOMS: Static-Stretching Application

Another area that is important in athletic populations is recovery and whether static stretching pre- or post-activity may assist in the recovery process and lead to a reduction in DOMS. Herbert et al.[37] reviewed the Cochrane Database and completed an updated review of the Cochrane Review, first published in 2007. They searched a number of listed articles and databases finding twelve viable studies to help assess and determine the effects of stretching before or after exercise on the development of DOMS. The evidence that they gathered from the randomized studies suggests that muscle stretching, whether conducted before, after, or before and after exercise, does not produce clinically important reductions in DOMS in healthy adults. Cheung et al.[38] also demonstrated that stretching had no effect on the alleviation of muscle soreness or other DOMS symptoms and that exercise is the most effective means of alleviating pain during DOMS.

There are many research papers with reference to static stretching. The summary below from just some of these papers provides some guidance in relation to static stretching from the references mentioned above.

- Ogura et al.[39] discussed the duration of static-stretching holds and concluded that short duration (30sec) of static stretching did not have a negative effect on muscle force production. Anything greater than 30sec holds may therefore reduce performance in relation to maximal strength output.
- Static stretching may compromise repeated sprint ability performance, but has less effect on change of direction speed performance (Beckett et al.[40]).
- A regular 10-week, 40min × 3 days each week static-stretching routine designed to stretch all the major muscle groups in the lower extremity may have positive gains on a number of performance variables for sedentary populations (Kokkonen et al.[41]).
- Static stretching before an endurance event may lower endurance performance, measured as distance covered, and it may increase the energy cost of running, demanding a higher calorific output (Wilson et al.[42]).
- After 30sec and 60sec of static stretching, flexibility measures may increase but peak torque may be reduced. Siatras et al.[43] therefore recommend that static-stretching exercises of a muscle group for more than 30sec of duration be avoided before performances requiring maximal strength.
- Acute bouts of static stretching for 45sec may negatively affect balance and reaction times (Behm et al.[44]).
- 30sec static stretches completed before high-intensity exercise lead to the time to complete a run being significantly less, and therefore quicker, in the non-stretch group than the stretching group (Lowery et al.[45]).
- Any change in ROM from static stretching for 30sec is short lived and will return to pre-stretching levels 3min after finishing the stretch (DePino et al.[46]).

- Muscle stretching, whether conducted before, after, or before and after exercise, does not produce clinically important reductions in DOMS in healthy adults (Herbert et al.[47]).

The above is a sample of many research papers and studies referencing static stretching and the pros and cons of different protocols. They hopefully provide an insight to help make appropriate applied decisions with athletes or clients.

Static-Stretching Application

Rationale: sedentary or older populations.

Outcome: may have positive gains on a number of performance variables for sedentary populations; no performance gains with acute application and even a reduction in performance measures (strength and power output).

Repetitions: >30sec holds.

Frequency: 3 × week.

Sets: two to three sets per body part.

DOMS Benefit: not relevant in static stretching.

Static stretching – lower body: quadriceps.

Hamstrings.

Glutes.

Adductors/groin.

Hip flexor.

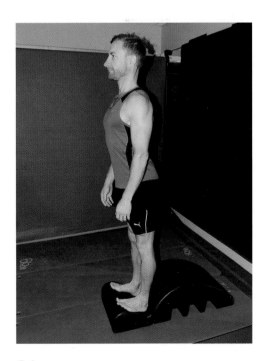

Calves.

Static stretching – upper body: triceps.

Chest/pectorals.

**Latissimus dorsi
(with Swiss ball).**

Latissimus dorsi (without Swiss ball).

Latissimus dorsi (standing).

DYNAMIC STRETCHING

Dynamic stretching consists of controlled leg or arm functional movement patterns that lead gently to the limits of a person's range of motion whilst moving. They can be completed passively, such as a dynamic calf stretch on a bench or on the floor, or actively, which would be when the hamstring muscle is moved, such as progressing the ROM in mid-air.

Dynamic stretching, in addition to leading up progressively to the muscle's functional ROM, has the added benefit of increasing body temperature, increasing heart rate and breathing rate in order to prepare the body for exercise. The different movement patterns will be completed whilst moving either on the spot, or over a certain distance, such as on a football or hockey pitch. This type of muscle preparation is typically used with athletes preparing to repeat similar movement patterns within their sport and who therefore need the body to be prepared for a full ROM activities.

Functional movement patterns that mimic the movements of the sport can be done within the warm-up and as part of dynamic stretching, for example hamstring flicks in order to warm up the hamstrings, which will also dynamically stretch the quadriceps muscles, or hamstring and groin movement patterns that may mimic a tackle or lunge in football or hockey. These movement patterns are very similar to the actual demands of the sport or game that the individual is preparing for. Heart rate will increase and the body will become well prepared for action. Historically, static stretching was the physical preparation

1. Start.

3. Full extension.

2. Extend.

4. Pull back to return to start position.

Dynamic flexibility: hamstrings.

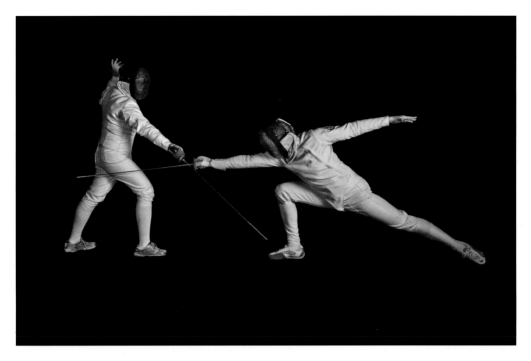

An attack in fencing demands a large range of movement.

of choice for athletes or sports teams; now it is dynamic stretching, also known as dynamic flexibility. But does this form of preparation optimize performance?

Some studies have looked at the combination of static stretching with a dynamic sport-specific warm-up. Samson et al.[48] discuss the effects of dynamic and static stretching within general and activity-specific warm-up protocols. Nine male and ten female subjects were tested under four warm-up conditions including: general aerobic warm-up with static stretching; general aerobic warm-up with dynamic stretching; general and specific warm-up with static stretching; and general and specific warm-up with dynamic stretching. Following all conditions, subjects were tested for: movement time (kicking movement of leg over 0.5m distance); counter-movement jump height; sit and reach flexibility; and six repetitions of 20m sprints. Results indicated that

when a sport-specific warm-up was included, there was a 0.94 per cent improvement in 20m sprint time with both the dynamic and static stretch groups. No such difference in sprint performance between dynamic and static stretch groups existed in the absence of the sport-specific warm-up.

The conclusions imply that completing a sport-specific warm-up in unison with either static stretching or dynamic stretching will improve performance. The static-stretch condition increased sit and reach ROM by 2.8 per cent over the dynamic condition. These results would support the use of static stretching within an activity-specific warm-up to ensure maximal ROM if that was needed for the sport, along with an enhancement in sprint performance.

Mascarin et al.[49] looked at whether stretch-induced reductions in throwing performance were attenuated by warm-up before exercise.

As we have discussed above in regard to isolated static stretching, it has been demonstrated that static stretching alone before exercise may lead to a reduction in muscular performance. Mascarin et al. looked at the effects of static shoulder-stretching exercises, dynamic warm-up exercises, or both together, on muscular performance evaluated by ball throwing. They used twenty-one female handball players and after completing one of the preparatory warm-ups, that being static stretching, dynamic stretching, or both, on the upper body, they completed both medicine ball throwing for distance and handball ball throwing for speed tests.

The study showed that static stretching performed prior to the medicine ball throwing test, a power test, reduced performance when compared with the dynamic warm-up exercises and when a dynamic warm-up exercise routine was added to the static stretching, the detrimental effects of static stretching on performance were negated and performance restored. The throwing speed was the same for all warm-up protocols. Mascarin et al. therefore recommend that athletes perform dynamic warm-up exercises together with static stretching prior to activity to avoid detrimental effects on muscle strength.

This combination of static stretching and dynamic warm-up could be a valuable tool for athletes whose role demands an increased ROM, for example a gymnast or an ice hockey goalkeeper. For others, it could be argued that there is a limited, or in fact no, role for static stretching in the warm-up and purely dynamic flexibility only should be completed. This would still involve a small-duration static stretch, of maybe 2–5sec, but would certainly avoid the long 20–30–40sec static stretches of years gone by.

Zourdos et al.[50] completed a repeat of the study of static stretching carried out by Wilson et al.[51] in 2010, which looked at the effects of static stretching on running energy cost and endurance performance. Zourdos et al. completed an identical protocol in 2012, but instead of using static stretching they used dynamic flexibility. They examined the effects of dynamic stretching on running energy cost and endurance performance in trained male runners.

Fourteen male runners performed a 30min preload run at 65 per cent VO_2max and a 30min time trial to assess running energy cost and endurance performance. The subjects either completed 15min of quiet sitting or 15min of dynamic stretching prior to the run. The results showed that caloric expenditure was significantly higher during the 30min preload run for the dynamic-stretching group (416.3 ± 44.9kcal), compared to the quiet sitting group (399.3 ± 50.4kcal). Contrary to the static-stretching protocol (Wilson et al.,[52]), there was little difference in the distance covered after quiet sitting (6.3 ± 1.1km), compared to the dynamic-stretching condition (6.1 ± 1.3km). These findings suggest that dynamic stretching does not affect running endurance performance in trained male runners and does not affect distance covered. So, why, if performance is equal to or even worse with the inclusion of a given mode of stretching, do we complete stretching and, specifically, dynamic-stretching exercises?

There are a number of reasons why we warm up prior to an event or exercise session, but mainly it is the perception that warming up will reduce the incidence of injury and will optimize performance by preparing the body for activity. An increase in body temperature, heart rate, breathing rate, speed of nervous impulses, sensitivity of nerve receptors and muscle blood flow, combined with a reduction in muscle viscosity, will all lead to improved preparedness before exercise. If static stretching has been shown to demonstrate reduced or impaired performance, while dynamic stretching has been shown to have no effect on distance covered, which is a performance measure, then why are we preparing this way? We need to ensure that the warm-up of choice is of benefit and will optimize performance.

Behm and Chaouachi[53] completed a review of the acute effects of static and dynamic stretching on performance. They concluded that a warm-up to minimize impairments and enhance performance should be composed of a submaximal intensity aerobic activity followed by large amplitude dynamic stretching and then completed with sport-specific dynamic activities. Applied simply, this would be a jog, perhaps with some curved runs, side slides, various multidirectional functional sport-specific movement patterns, followed by some dynamic flexibility exercises and progressing on to more explosive sport-specific demands.

If we consider injury prevention and how we warm up based on the above criteria, how can we determine if the stretching component of the warm-up is indeed preventing any injury? We could quite easily presume that injury prevention results from the preparedness of the body. I believe that working a warm muscle through a ROM will appropriately prepare the body for those specific movement patterns during competition or training, but certainly increasingly body temperature and being physiologically prepared appears to be more important relating to athletic preparedness and injury prevention than any type of stretching.

Witvrouw et al.[54] look at the misconceptions and conflicting research in relation to stretching. They discuss the stretch-shortening cycle and how different sports demand and require a muscle-tendon unit that is compliant enough to store and release the high amount of elastic energy that

benefits performance. They discuss whether the structure can cope with the demands of the sport and highlight that recent studies have shown that stretching programmes can influence the viscosity of the tendon and make it significantly more compliant, and when a sport demands stretch-shortening cycles of high intensity, stretching may be important for injury prevention. In contrast, when the type of sports activity contains low-intensity, or limited, stretch-shortening cycles (for example, jogging, cycling and swimming), there is no need for a very compliant muscle-tendon unit and therefore no requirement for pre-event stretching. This is an interesting concept, which may be very relevant to the millions of recreational runners out there who may in fact have no physiological requirement to stretch prior to their run.

There is some debate as to the optimal level of flexibility required to aid performance and prevent injury. It appears to be desirable to have a functional level of flexibility, with research reporting that individuals with both extremes of flexibility, these being decreased ROM or stiffness and hypermobility, appear to have a greater risk of injury than the average group.

Perrier et al.[55] compared the effects of a warm-up with static versus dynamic stretching on counter-movement jump (CMJ) height, reaction time and low back and hamstring flexibility (sit and reach test). They wanted to determine whether any observed performance deficits would occur with each treatment. Twenty-one recreationally active males completed the research and each session included a 5min treadmill jog followed by one of the stretch treatments: no stretching; static stretching; or dynamic stretching. The researchers discovered that the CMJ height was greater for the dynamic-stretching group (43.0cm), than for the no-stretching group (41.4cm) and the static-stretching group (41.9cm). The analysis of reaction time, which was determined from measured ground-reaction forces during the CMJ, showed no significant effect as a result of the treatments. Treatment, however, had a main effect on flexibility via the sit and reach test. Flexibility was greater after both static stretching and dynamic stretching compared to after no stretching, with no difference in flexibility between static stretching and dynamic stretching. Perrier et al. conclude that athletes in sports requiring lower-extremity power should use dynamic-stretching techniques when warming up to enhance flexibility while improving performance, as observed in the greater performance from the counter-movement jump tests.

Needham et al.[56] looked at elite youth soccer players and, similar to Perrier et al.'s study above, employed a 5min low-intensity jog followed by 10min of static stretching, dynamic stretching, or dynamic stretching followed by eight front squats + 20 per cent body mass (resistance). Subjects performed a counter-movement jump followed by a 10m and 20m sprint test immediately and at 3min and 6min after each warm-up protocol. By adding in the resistance prior to testing, the dynamic stretching and resistance group achieved better performance measures on the CMJ and also the 10m and 20m sprint tests at both 3min and 6min post-warm-up compared to the dynamic-stretching group, which in turn performed better than the static-stretching group. The results of the study suggest that a dynamic warm-up with the inclusion of resistance enhances jumping ability more than dynamic exercise alone. In addition, a dynamic warm-up produces a superior sprint and jump performance compared to a warm-up consisting of static stretching.

Yamaguchi et al.[57] looked at a comparison of static stretching for 30sec and dynamic-stretching comparisons on leg-extension power.

Leg-extension power was measured by a leg-extension power measurement system. Each subject performed the same modalities on random separate days – static stretching and dynamic stretching on the five muscle groups in the lower limbs and non-stretching on separate days. Leg-extension power was measured before and after the static-stretching, dynamic-stretching and non-stretching days. The non-stretching and static-stretching days both demonstrated no significant difference in leg-extension power before or after the static or non-stretching protocol (static stretching 1,788.5 +/−85.7Watts and non-stretching 1,784.8 +/−108.4Watts). However, the dynamic-stretching protocol demonstrated significantly greater power post-dynamic stretching (2,022.3 +/−121.0Watts). These results suggest that static stretching for 30sec neither improves nor reduces muscular performance and that dynamic stretching enhances muscular performance.

The difference between dynamic stretching and ballistic stretching is the force applied at the end ROM, going beyond the normal range in ballistic stretching. With dynamic stretching being a perceptively safer option than ballistic stretching by the nature of the movement, it may be more appropriate to complete dynamic stretching over ballistic stretching as a 'dynamic' stretch of choice. Ballistic stretching is discussed in more depth below.

It appears that overwhelmingly dynamic stretching is preferential to static stretching where performance is concerned and therefore should be the stretching mode of choice for athletes. It has been shown to improve the ROM in addition to improved performance measures within a number of studies. It is interesting to note the benefit observed with the research from Needham et al.[58] with the inclusion of resistance jumps within the warm-up prior to explosive activities and how this may also enhance performance.

Dynamic-Stretching Application

Rationale: warm-up prior to a sporting event or before physical activity.

Outcome: enhances muscular performance and flexibility/ROM before exercise.

Repetitions: 20–30sec per body part.

Sets: one to two sets per body part.

Duration: 6–8min prior to exercise × 2 blocks if desired. Research indicates that one set appears to be enough for performance gains.

Frequency: prior to all training and sporting events.

DOMS benefit: not relevant in dynamic stretching.

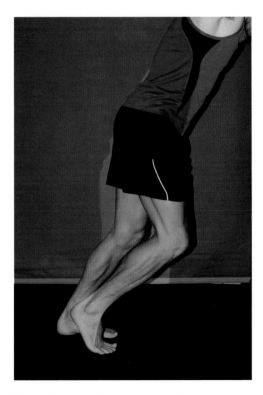

Dynamic stretching: calf muscles.

1. Right side.

2. Left side.

3. Right side.

4. Left side.

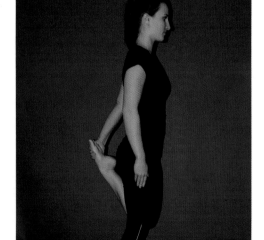

Quadriceps. Hold for 2–3 seconds, feeling a stretch sensation, then switch legs dynamically.

1. Start position.

2. Extension.

3. Full extension.

4. Return to start position and switch legs.

Hamstrings. Alternative left side and right side dynamically through range.

1. Left side.

2. Right side.

3. Left side.

4. Right side.

Glutes. Alternative left side and right side dynamically through range.

1. Start position.

2. Extension.

3. Full extension.

4. Return to start position and switch legs.

Adductors/groins. Alternative left side and right side dynamically through range.

1. Right side.

3. Right side.

2. Left side.

4. Left side.

Hip flexors. Alternative left side and right side dynamically through range.

BALLISTIC STRETCHING

Ballistic stretching is when the momentum of a moving body or a limb is used to force it beyond its normal range of motion. This is different to dynamic stretching, which gradually builds up to a full ROM, as described above. Bouncing movements to increase the range beyond the normal ROM are observed in ballistic stretching and concerns over its safety have been documented for this reason. When a movement is completed as quickly and explosively as possible, this could be defined as ballistic. Ballistic actions are characterized by high firing rates, brief contraction times and high rates of force development. Appropriate application of ballistic movement patterns may include sports like karate or basketball, for example, where explosive movement is essential.

Bacurau et al.[59] investigated the acute effect of a ballistic-stretching and a static-stretching exercise protocol on flexibility and lower-limb maximal strength. Fourteen physically active women performed three experimental sessions: a control session (estimation of 45 degrees leg press, one repetition maximum [1RM]); a ballistic session (20min of ballistic stretch and 45 degrees leg press 1RM); and a static session (20min of static stretch and 45 degrees leg press 1RM). The results indicated that maximal strength decreased after static stretching, but it was unaffected by ballistic stretching. In addition, static-stretching exercises produced a greater acute improvement in flexibility compared with ballistic-stretching exercises. Static stretching may not be recommended before athletic events or physical activities that require high levels of force, but we know that it increases acute ROM measures. However, ballistic stretching could be more appropriate because it seems less likely to decrease maximal strength.

Jaggers et al.[60] looked at the acute effects of dynamic and ballistic stretching on vertical jump height, force and power. Considering the anecdotal concerns about injury risk associated with ballistic stretching, it is good to compare both stretching modalities within the same research paper in relation to performance measures and then to consider the appropriate application for each stretching mode. The purpose of their study was to compare the differences between two sets of ballistic stretching and two sets of a dynamic-stretching routine on vertical jump performance. All subjects completed three individual testing sessions on three non-consecutive days. On each day, the subjects completed one of three treatments: no stretch; ballistic stretch; and dynamic stretch.

The ballistic-stretching protocol consisted of the following five ballistic stretches targeting muscles used during a counter-movement jump: forward lunge; supine knee flex; sitting toe touch; quadriceps stretch; and the butterfly. Each subject was instructed to perform the ballistic stretches by bouncing rapidly for 30sec at a pace of 126 beats per minute maintained by a metronome. Each subject performed two sets of the stretches. A technician timed each subject for a full 30sec stretch.

The dynamic-stretching protocol consisted of five stretches that targeted the same muscle area that was stretched in the ballistic routine. Those stretches were: leg kickbacks; standing knee raise; calf raise; hurdle step-overs; and butt kicks. These stretches were performed five times, slowly at first, then ten times as quickly and powerfully as possible without bouncing, for a total of fifteen repetitions. Each subject completed two sets of each stretch.

The results indicate that there was no significant difference when comparing no stretch with ballistic stretch for jump height, force, or power. Similarly, there were no significant differences when comparing no stretch with dynamic stretch for jump height or force. However, this investigation did find a significant increase in jump power when comparing no stretch with dynamic stretch. So according to this study by Jaggers et al. there is no performance benefit from completing ballistic stretching prior to exercise for power, jump height or force production, having the same outcome as no stretching. Dynamic flexibility, however, did demonstrate a positive outcome in jump power.

Woolstenhulme et al.[61] reviewed the effect of four different warm-up protocols followed by 20min of basketball activity on flexibility and vertical jump height. Subjects participated for six weeks, two times per week, of warm-up and basketball activity. The warm-up groups participated in ballistic stretching, static stretching, sprinting or basketball shooting (which was the control group). The static stretches and ballistic stretches were the same, comprising the same four stretches, but performed relative to the appropriate stretching mode. Static stretches were held for 30sec × 2 on each stretch, with 15sec rest between each stretch. The ballistic stretches were the same stretches as the static stretches, but were completed at end ROM, bouncing to a 60sec metronome beat for the 30sec duration, again with 15sec rest between sets.

Woolstenhulme et al. measured sit and reach and vertical jump height before (week one) and after the six weeks (week seven). The group posed three questions:

- What effect does six weeks of warm-up exercise and basketball play have on both flexibility and vertical jump height?
- What is the acute effect of each warm-up on vertical jump height?
- What is the acute effect of each warm-up on vertical jump height following 20min of basketball play?

Firstly, Woolstenhulme et al. established that flexibility increased for the ballistic, static and sprint groups compared to the control group, with the biggest increase in sit and reach score from the ballistic group, while vertical jump height did not change for any of the groups. Secondly, they measured vertical jump immediately after the warm-up on four separate occasions during the six weeks (at weeks zero, two, four and six), concluding that vertical jump height was not different for any group. Finally, they concluded that only the ballistic stretching group demonstrated an acute increase in vertical jump 20min after basketball play. Vertical jump height was measured immediately following 20min of basketball play at weeks zero, two, four and six.

The results imply that it would be beneficial for athletes to complete ballistic stretching as

1. Start position.

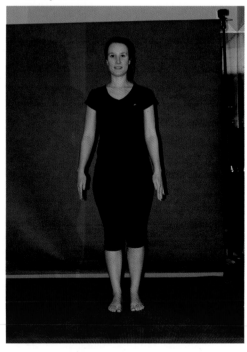

Ballistic stretching: hamstring stretch (above and opposite).

2. Extension.

3. Further extension.

4. Further extension.

5. Full range ballistic extension.

a warm-up for basketball play, as it aids vertical jump performance, which is a potential key performance indicator in basketball. There was an increase of around 3cm in each of the ballistic stretching group after the six-week analysis, which is a significant improvement. Much more research is needed to recognize the appropriate application of ballistic stretching, but for a stretching mode that has a bad reputation, it may be able to enhance performance significantly.

Woolstenhulme *et al.* also discuss the statements from many other research papers claiming that ballistic stretching may increase injury risk or DOMS, but imply that these statements are incorrect and misleading. The movement patterns completed within ballistic stretching are very similar to the explosive nature of basketball and martial arts, for example, therefore the sport-specific nature of ballistic stretching may well prepare the athlete for performance in an appropriate manner. The application is very similar to dynamic stretching, just beyond range, and if an individual is conditioned to complete such movement patterns, it appears that ballistic stretching may well be an appropriate sport-specific training modality with performance gains.

Ballistic-Stretching Application

Rationale: sport-explosive warm-up prior to a sporting event or before physical activity that demands ballistic/explosive movements.

Outcome: increased ROM/flexibility; improved vertical jump.

Repetitions: 30sec duration on each side for each limb.

Sets: two sets on all selected muscle groups.

Rate/frequency of stretch: metronome beat around sixty beats/minute.

Frequency: before sports events and before training so as to be conditioned for explosive movement patterns.

DOMS benefit: not relevant in ballistic stretching.

PNF AND ISOMETRIC STRETCHING

The application of proprioceptive neuromuscular facilitation (PNF) involves the use of a muscle contraction before the stretch in an attempt to achieve maximum muscle relaxation and a reciprocal increased stretch response. It is a technique first named and developed by Dr Herman Kabat in the early 1940s as proprioceptive facilitation. In 1954, Dorothy Voss added the word 'neuromuscular' to give us the now familiar concept of PNF.

Dr Kabat's development of PNF came from his experience as a neurophysiologist and physician and his work with polio patients, whom his team helped with specific stretching and strengthening activities. Kabat integrated manual techniques inspired by Charles Sherrington's discovery of successive induction, reciprocal innervation and muscular inhibition. Sherrington was a pioneer in neurophysiology, winning the Nobel Prize in Physiology or Medicine in 1932 for work related to neurons. Sherrington received the prize for showing that reflexes require integrated activation and demonstrated reciprocal innervation of muscles. He established two Laws that changed the understanding of how the human body innervates:

- **First Law:** every posterior spinal nerve root supplies a particular area of the skin, with a certain overlap of adjacent dermatomes.
- **Second Law:** the law of reciprocal innervation, in which contraction of a muscle

PNF: self-managed PNF with a strap.

1.
Resist against the band.

2.
Increase ROM.

3.
Resist against the band.

4.
Increase ROM.

is stimulated and there is a simultaneous inhibition of its antagonist. It is essential for coordinated movement.

Following on from Sherrington's Laws, the application of PNF involves the use of a muscle contraction before the stretch in an attempt to achieve maximum muscle relaxation and an increased stretch response through the ROM. It is a partner-assisted exercise, or can be self-managed with the use of a strap or band, for example. It can be used to increase the ROM of the muscle significantly by resisting, or by pushing against resistance, then relaxing the muscle into an increased ROM through the muscle's resultant inhibition.

There is some confusion within the literature regarding the actual mechanism of increased ROM in the specific muscle that has been stimulated as a result of the reciprocal inhibition from the antagonist muscle. As Sharman et al.[62] discuss, the superior changes in ROM that PNF stretching often produces compared with other stretching techniques has traditionally been attributed to autogenic and/or reciprocal inhibition, although, Sharman claims, the literature does not support this hypothesis. Instead, and in the absence of a biomechanical explanation, the contemporary view proposes that PNF stretching influences the point at which a stretch is perceived or tolerated. The mechanism(s) underpinning the change in stretch perception or tolerance are not known, although pain modulation has been suggested.

Chalmers[63] also re-examines the possible role of Golgi tendon organ and muscle spindle reflexes in PNF muscle stretching. Previously, it was thought that the influence of Golgi tendon organs on the stretch receptors would affect the stretch reflex within the muscle.

DEFINITION OF GOLGI TENDON ORGANS

Golgi tendon organs are receptors located in muscle tendons that provide mechanosensory information to the central nervous system about muscle tension.
(Purves, D. et al., Neuroscience, 2nd Edition [Sunderland, MA: Sinauer Associates; 2001].)

Chalmers concludes that changes in the ability to tolerate stretch and/or the viscoelastic properties of the stretched muscle are possible mechanisms that allow for an increased ROM.

Higgs et al.[64] evaluated possible strength changes as a result of a stretching modality, as we know that static stretching can negatively affect power output. They also looked at the effect of a four-week PNF stretching programme on isokinetic torque production. The range of motion was recorded before and after the first stretching session of each week. At the end of a four-week period, the peak isokinetic quadriceps torque and flexibility were again measured. They discovered that in a four-week quadriceps flexibility training programme consisting of three cycles of PNF stretching performed three times a week, the mean improvement in the knee flexion ROM over the whole programme was 9.2 degrees and typical gains after a single stretching session were around 3 degrees. The peak isokinetic torque produced at 120 degrees or at 270 degrees were unchanged.

These findings suggest that it is possible to improve flexibility using three PNF stretch cycles performed three times a week without altering muscle isokinetic strength characteristics. This is an important consideration, especially for post-operative individuals, for example, when they may need to increase the ROM of a joint, but not at a consequence of a reduction in strength. It has also been demonstrated in the study by O'Hora et al.,[5] where the aim was to compare the effectiveness of a single bout of a therapist-applied 30sec static stretching versus a single bout of therapist-applied 6sec hamstring (agonist) contract PNF.

Forty-five subjects were randomly allocated static stretching, PNF stretching or control (no stretching). The flexibility of the hamstring was determined by a range of passive knee extension, measured using a universal goniometer, with the subject in the supine position and the hip at 90 degrees flexion, before and after intervention. Both stretching modes led to greater ROM compared to the control group, static stretching increasing by 7.53 degrees and PNF by 11.80 degrees. The PNF group demonstrated significantly greater gains in knee extension compared to the static-stretching group, with a mean difference of 4.27 degrees after one bout of treatment. The protocol did not include a warm-up and the longevity of the increased ROM within the hamstrings is unknown, but if an increase in ROM is desired then a single bout of 6sec PNF has been shown to significantly increase this.

So, what level of force should be applied to resist the limb when an increased range is desired? Research by Kwak and Ryu[66] examined the effect of contraction intensity on ROM while applying PNF stretching. Three contraction intensities of 100 per cent, 60 per cent and 20 per cent plus a control group were assessed in the study, with optimal ROM achieved at 100 per cent and 60 per cent contraction intensity. So when using PNF as a stretching modality, it is important to consider contraction intensity with the individual. Even at 20 per cent contraction intensity there was more increased range than with the control group. This is important information, especially if PNF is being applied to injured or special populations. Isometric contractions, when the limb does not move during resistance, or when load is applied, is an important training modality for early rehabilitation, so PNF surely has a role within rehabilitation and it appears to be a safe modality that will achieve increases in ROM without any detrimental effect on performance.

PNF stretching can be completed as follows:

1. Move into the stretch position so that you feel the stretch sensation.
2. Your partner holds the limb in this stretched position.
3. You then push against your partner by contracting the antagonistic muscles for 6–10sec, then relax. During the contraction, your partner aims to resist any movement of the limb (isometric hold).
4. Your partner then moves the limb further into the stretch until you feel the stretch sensation.
5. Go back to step 2 with the isometric hold. (Repeat this procedure three or four times before the stretch is released.)

PNF Application

Rationale: rehabilitation setting or when increased ROM is desired.
Outcome: improved ROM with no subsequent effect on strength or power.
Repetitions: three to four reps per limb per set.
Sets: one to two sets per limb.
Duration: 6–10sec isometric holds per repetition.

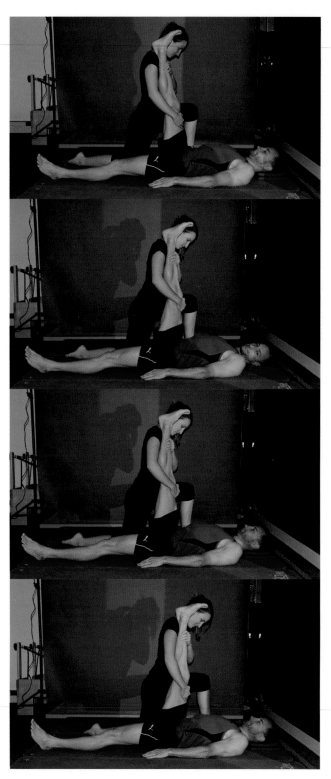

Partner-assisted PNF: hamstring stretch.

1. and 2.
Move into a stretch position with your partner holding the limb.

3.
Push against your partner who resists against the limb force (isometric hold).

4.
The partner moves the limb further to elicit a stretch sensation.

5.
Repeat again until desired ROM achieved.

Contraction intensity: between 20–100 per cent.; optimal = between 60–100 per cent.

Frequency: 3 × each week.

DOMS benefit: not relevant in PNF/isometric stretching.

MYOFASCIAL RELEASE

Myofascial release is a hands-on technique that provides sustained pressure into myofascial restrictions within the muscle to eliminate pain and restore movement. Fascia is strong connective tissue that performs a number of functions, including enveloping and isolating the muscles of the body, providing structural support and protection.

Underneath the superficial fascia lies deep fascia, a much more densely packed and strong layer of fascia. Deep fascia covers the muscles in connective tissue, which help to keep the muscles divided and protected. On occasion, this fascia can create tight knots or connective adhesions, which act as trigger points that can cause pain.

Myofascial release is a type of soft-tissue massage that incorporates stretching and massage of the connective tissues or fascia. A tennis ball or a foam roller may be used, as well as manual therapy, to place pressure on

Fascia in the human body: shown in the white bands across the muscles.

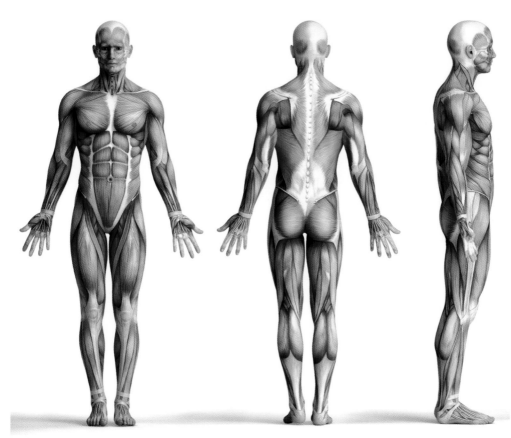

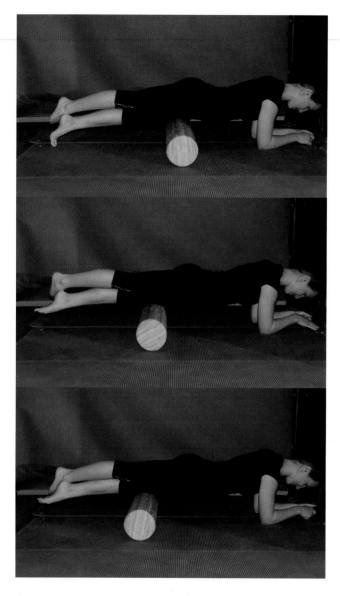

Myofascial release with a foam roller: quadriceps. Slowly move up and down the muscle of the chosen limb.

the affected local area within the muscles and this process can be self-administered, which is of course an advantage.

According to practitioners of myofascial release, poor posture, injury, illness or stress can negatively affect body alignment and cause fascia to become restricted. They believe that this can cause pain and impair movement. A gentle form of stretching and manual compression is said to restore flexibility to this connective tissue and provide relief from fascial restrictions and pain. Myofascial release aims to access these trigger points, thereby freeing up the muscle and allowing it to move more easily and effectively. Trigger points have been defined as areas of muscle that are painful to palpation and are characterized by the presence of what seem like knots within the muscle.

But does it work? Grieve et al.[67] looked at

DEFINITION OF TRIGGER POINTS

A sensitive area of the body, stimulation or irritation of which causes a specific effect in another part, especially a tender area in a muscle that causes generalized musculoskeletal pain when overstimulated.

(Oxford English Dictionary [Oxford: OUP, 2015].)

the immediate effect of bilateral self-myofascial release on the plantar surface of the feet and its effect on hamstring and lumbar spine flexibility. The study raised the question that there was currently no evidence to support the effect of bilateral self-myofascial release from the plantar surface of the foot and a possible resultant positive effect on flexibility of lumbar spine and hamstring flexibility upwards through the kinetic chain. Baseline and post-intervention flexibility were assessed via the sit and reach test. What they found was that after one acute treatment of 2min of seated treatment per foot (a total of 4min), the intervention group significantly increased their sit and reach scores compared to the control group. This may lead to possible positive outcomes for future clinical practice and may be an effective variation to help with stretching and mobility issues for special populations. The study does not indicate for how long the benefits last, but certainly there was a significant acute change in flexibility as a result of plantar surface self-myofascial release.

Changes in hamstring flexibility range of motion in relation to foam roller application was studied by Couture *et al.*[68]. They looked at the effect of foam-rolling duration on hamstring flexibility. The knee-extension range of

motion of thirty-three college-aged men and women was assessed after a short (two sets of 10sec) and long (four sets of 30sec) duration of hamstring self-administered myofascial release using a commercial foam roller. They discovered that there was no change from baseline measures with the test durations, so concluded that a total duration of up to 2min is not adequate to induce improvements in knee-joint flexibility. Longer duration of treatment appears to be necessary for any improvement in ROM.

We know that PNF has positive effects on flexibility and ROM, so Junker *et al.*[69] compared PNF stretching with myofascial release and a control group. Hamstring flexibility was measured by a stand and reach test before and after the intervention period. PNF was assigned to twelve sessions of contract – relax PNF stretching in a four-week block. The control group had no intervention and the self-myofascial release group massaged their hamstring muscles with the foam roller three times per week for four weeks, totalling twelve total training sessions.

The results indicated that both the foam roller and the PNF technique improved hamstring ROM as measured on the stand and reach test and Junker *et al.* conclude that the foam roller can be seen as an effective tool to increase hamstring flexibility within four weeks. The effects are comparable with the scientifically proven contract – relax PNF stretching method according to their study and we know from our discussion about PNF that it causes a measureable increase in joint ROM when applied.

We have discovered that there is a negative physical outcome with a few types of stretching techniques, for example that there is a loss in muscular performance with static stretching. MacDonald *et al.*[70] looked at determining the effect of self-myofascial release via foam-roller application on knee-extensor force and activation and knee-joint ROM. They measured quadriceps maximum voluntary contraction force, evoked force and activation, plus knee-joint ROM for eleven subjects. These components were measured before, 2min and 10min after two conditions: 2 × 1min trials of self-myofascial release of the quadriceps via a foam roller; and no self-myofascial release (control group). There were no significant differences between conditions for any of the neuromuscular-dependent variables, that being quadriceps maximum voluntary contraction force, evoked force and activation. However, after foam rolling, subjects' ROM significantly increased by 10 degrees and 8 degrees at 2min and 10min, respectively. MacDonald et al. therefore conclude that an acute bout of self-myofascial release of the quadriceps is an effective treatment to enhance knee-joint ROM acutely without a concomitant deficit in muscle performance.

This is important for the application of self-myofascial release and when to administer it. It can be concluded, although much more research is needed, that it would be appropriate to complete self-myofascial release prior to exercise or activity if increased ROM is desired without any subsequent reduction in muscular force.

Healey et al.[71] determined that there was no effect on performance measures as a result of foam rolling as they looked at the effects of myofascial release with foam rolling on performance. The physical tests that their subjects completed were measures in vertical jump height and power, isometric force and agility. All four tests had no significant improvement or reduction in performance as a result of self-myofascial release, but post-exercise fatigue after foam rolling was significantly less than it was in the control group. The reduced feeling of fatigue as a result of self-myofascial release may allow participants to extend acute workout time and training

volume, which can lead to long-term performance enhancements and opportunity to train in a recovered state. Importantly for athletic populations, it was concluded that foam rolling had no effect on performance, but had positive benefits on post-exercise fatigue.

DOMS and fatigue can be a limiting factor affecting an ability to perform repeated bouts of high-intensity exercise during certain high-intensity or unaccustomed exercise training phases.

DOMS: Myofascial Release Application

DOMS is a desirable outcome from an intense training load, but certainly at the highest level an athlete may want to maintain or extend training load, intensity and volume on a daily basis to stimulate a positive training effect and to limit the debilitating feeling of DOMS and enhance recovery. Self-myofascial release may be an effective modality to assist in the management of DOMS and improve recovery. Indeed, Pearcey et al.[72] looked at foam rolling for DOMS and recovery of dynamic performance measures.

They examined the effects of foam rolling as a recovery tool after an intense exercise protocol through the assessment of pressure-pain threshold, sprint speed (30m sprint time), power (broad-jump distance), change of direction speed (T-test) and dynamic strength-endurance (barbell back squats at 70 per cent 1RM). The participants in this study performed two conditions, separated by four weeks, involving ten sets of ten repetitions of back squats at 60 per cent of their 1RM, with a focus on eccentric loading to increase DOMS, followed by either no foam rolling or 20min of foam rolling immediately, 24hr and 48hr post-exercise. The protocol for the foam roller was prescribed as follows:

- As much muscle mass as tolerable placed on to the roller, starting at the distal end of the muscle.
- Rolling the body back and forth along the roller.
- The cadence of movement was prescribed as fifty beats per minute (move every 1.2sec).
- This was performed for 45sec with a 15sec rest between each muscle group.
- This was completed once per lower extremity muscle group, left and right side and took around 20min – 15min of rolling and 5min of rest.
- The muscles addressed were: quadriceps, adductors, hamstrings, iliotibial band and gluteals.
- This was completed immediately, 24hr and 48hr post-exercise.

Pearcey et al. concluded that DOMS negatively affected all performance measures when no foam rolling was performed. Once the group had completed the foam-rolling protocol, it had a moderate effect on an increase in sprint time and a small effect on the decline in the broad jump. Change of direction was less affected and it had a moderate effect on strength endurance. The pressure-pain threshold was reduced when foam rolling was completed, with a large decrease in the pressure-pain threshold at 24hr and 48hr and only a small decrease at 72hr. The 24–48hr benefit is important for athletes, as training will usually be scheduled within those time frames, often training on consecutive days.

Schroeder and Best[73] completed a literary review on self-myofascial release and its use for pre-exercise, recovery or maintenance. They concluded that the evidence is such that self-myofascial release appears to have a positive effect on ROM and soreness/fatigue following exercise. They have pointed at the requirement for further study to help define optimal parameters (timing and duration of use) to aid performance and recovery. Whist we await further research from the reviews, an effective application of self-myofascial release may be as follows:

- Move the foam roller or tennis ball around the particular body part until you feel some discomfort. This is the trigger point.
- Hold the foam roller or tennis ball there until the muscle relaxes; this may take up to 45sec, or until the muscle has relaxed and the pain has gone.
- Resume rolling until you find another trigger point.
- Address all the required muscle groups.
- Stop and hold again until the pain goes.
- A treatment can take as long as necessary, but usually between 10–15min is sufficient.

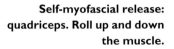

Self-myofascial release: quadriceps. Roll up and down the muscle.

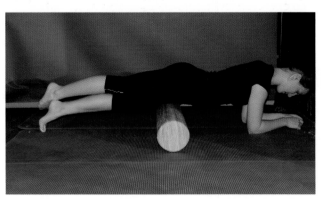

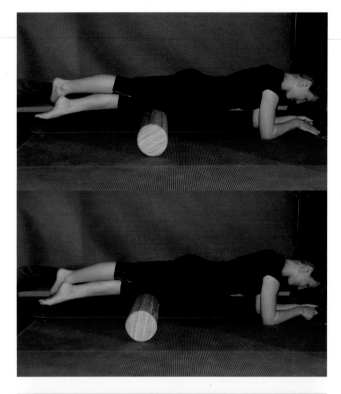

Quadriceps (cont).

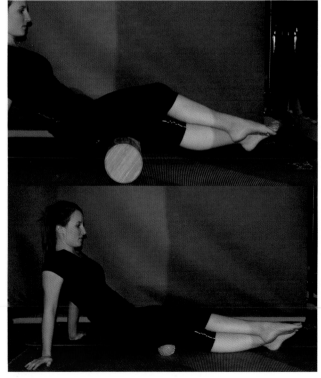

Hamstring.

Foam roller hamstring.

Tennis ball or spiky ball hamstring.

<u>*Calf muscles.*</u>

Foam roller calf.

Spiky ball calf

Tennis ball calf.

<u>*Gluteals.*</u>

Spiky ball gluteals.

Tennis ball gluteals.

Adductors.

Foam roller adductors.

Iliotibial band (ITB).

Foam roller (ITB).

Spiky ball ITB.

Myofascial Release Application

Rationale: recovery from DOMS pre- and post-exercise application for increased ROM with no subsequent performance losses.

Outcome: improved recovery with reduction in pain from DOMS, increased ROM and also improved performance indicators post-high-intensity exercise.

Repetitions: one to two repetitions per muscle group.

Movement cadence: every 1sec or so on an area working up and down a particular muscle group.

Duration: 45sec moving up and down the muscle, per muscle group.

Rest: 15sec rest between each muscle group.

Frequency: pre-exercise for improved ROM and immediately post-exercise and 24hr, 48hr after intense sessions to help reduce DOMS.

DOMS application: positive outcome with myofascial release.

Vibration platform.

WHOLE-BODY VIBRATION TRAINING AND STRETCHING

Vibration training hit gym facilities in earnest about ten to fifteen years ago. The on-going quest for the latest fitness craze and novelty that would make a difference in the physical training of millions was the goal for manufacturers, although the concept of whole-body vibration training had been around for many years before. Both Russian and NASA scientists have completed research on vibration training, the Russians for elite athletes, NASA for astronauts in space who experience bone density loss due to zero gravity. But can whole-body vibration training help with improved levels of flexibility?

Vibration training requires the use of specially designed machines that vibrate at specific frequencies, which are normally between 30Hz and 50Hz. They are platform-based machines and can vary by having a tilt platform or a non-tilt platform. A variety of exercises can be completed while standing, sitting, or by placing your hands or lower limbs on the vibrating plate to perform upper body or lower body exercises, such as triceps dips, lunges or press-ups.

The physiology of vibration training activates and excites all of the muscle fibres during a treatment or training session, which in turn ensures that the muscles are all contracting at incredibly high frequencies and being subjected to considerable forces. Vibration is delivered in different frequencies and is measured by Hertz. The vibration will be

DEFINITION OF VIBRATION TRAINING

The concept and physiology of vibration training and what happens to the body is quite complex. However, an effective explanation is provided by D.J. Cochrane and S.R. Stannard in their paper 'Acute Whole Body Vibration Training Increases Vertical Jump and Flexibility Performance in Elite Female Field Hockey Players' (*British Journal of Sports Medicine*, 2005, 39, pp. 860–5):

> *Vibration enhances the stretch reflex loop through the activation of the primary endings of the muscle spindle, which influences agonist muscle contraction while antagonists are simultaneously inhibited. The enhanced flexibility measure following whole body vibration suggests that vibration exposure may have activated the inhibitory interneurones of the antagonist muscle. This in turn may have caused changes to intramuscular coordination to decrease the braking force around the hip and lower back joints and potentiate the sit and reach score [in their research]. Increases in static and dynamic muscular contractions have been attributed to muscle stiffness, which is a function of muscle and tendon components. The magnitude of the stretch load and the condition of the musculotendinous complex ultimately determine which reflexes dominate. For pre-stretching to enhance concentric muscular contraction, excitatory responses of the muscle spindle must exceed the inhibitory effects of the Golgi tendon organ (GTO). This is normally achieved through potentiated neural input of muscle spindle sensitivity or suppression of GTO neural activity. In strength and power training, performing heavy sets of squats has been shown to augment jump squat height. Equally in whole body vibration, fast joint rotation and muscle stretching occur, which is likely to increase muscle stiffness following the purported neural potentiation of the stretch reflex pathway and ∞ motor neurone input. Moreover, vibration causes more excitatory responses to the primary endings of muscle spindles compared to secondary endings and GTOs. The joint, skin, and secondary endings also detect the vibratory stimulus whereby the neural activity of the primary endings is potentiated through the activation of the γ motor neurone. This post-activation potentiation may explain the concomitant enhancement of flexibility performance.*

distributed in an oscillatory motion to determine the amplitude. The applied vibration will increase muscular blood flow and could therefore be an efficient recovery modality to help flush the muscles of post-exercise toxins. We also know that from dynamic-stretching research that an increase in blood flow within the muscles generally causes an increase in muscular performance, so this may well also affect the outcome of vibration training before exercise.

Dallas et al.[74] examined the acute effects of different vibration loads (frequency and amplitude) of whole-body vibration on flexibility and explosive strength of lower limbs in springboard divers. All eighteen volunteers completed three different whole-body vibration protocols. Flexibility (sit and reach test) and explosive strength of lower limbs: Sargent jump, counter-movement jump, single leg jump for right leg (RL) and left leg (LL) were measured before, immediately after and 15min after the end of vibration exposure.

The three protocols with different frequencies and amplitudes were used in the present study for a total duration of 2min per protocol (4 × 30sec with 30sec rest between sets). The test sets were either low vibration frequency and amplitude (30Hz/2mm), high vibration frequency and amplitude (50Hz/4 mm), or a control protocol with no vibration. Each athlete completed 4 × 30sec exercises each on the platform. The exercises included: static squat at a knee angle of 120 degrees; dynamic squat at a tempo of 2sec up and 2sec down at a knee angle ranging from 120 degrees to 180 degrees; and two lunges (one on each leg) with the working vibrated leg on the platform and the other leg on the ground.

The results indicated that flexibility and explosive strength of lower limbs were significantly higher in both whole-body vibration protocols compared to the no-vibration group. The greatest improvement in flexibility and explosive strength, which occurred immediately after vibration treatment, was maintained 15min later in both whole-body vibration protocols.

So can vibration training improve the ROM for athletes who require vast levels of flexibility, like a gymnast, for example? Sands *et al.*[75] looked at the use of vibration to enhance acute ROM, while assessing the influence of vibration and stretching on pressure to pain threshold perception. Stretching beyond the ROM of a limb is often prescribed by coaches in performance sports that demand full functional range, like gymnastics, ice skating or even a field hockey or an ice hockey goalkeeper.

Sands looked at ten young male gymnasts and assessed them for split ROM. One side split was randomly assigned as the experimental condition, while the other side split was assigned as the control. Both side splits

were performed on a vibration device: the experimental condition had the device turned on at 30Hz for 45sec and the control condition was performed with the device turned off. Pressure to pain threshold was also measured via an algometer on the biceps femoris and vastus lateralis. The subjects were asked to identify when the pressure placed upon the two areas changed from a feeling of pressure to a feeling of pain. Algometer force values recorded pre- and post-vibration showed no change in pain perception as a result of vibration exposure. Pre- and post-flexibility difference scores were determined between the vibrated split and the non-vibrated split by measuring the measured height of the anterior superior iliac spine from the floor using a metre stick and showed a statistically significant difference of 2.3cm to 5.8cm post-vibration, indicating that vibration and stretching increased ROM more than stretching alone.

Fagnani *et al.*[76] investigated the short-term effects of an eight-week whole-body vibration protocol on muscle performance and flexibility in female competitive athletes. Twenty-four subjects completed the study and were split into non-vibration (control group) and vibration groups. The vibration intervention consisted of an eight-week whole-body vibration at 35Hz, three times a week, employed by standing on a vertical vibration platform. Two exercises were completed during the vibration training:

- **Position one:** 90 degrees double leg squat with hands on hips.
- **Position two:** single leg 90 degrees squat with hands on hips and non-standing leg in the air.

Duration of vibration varied over the eight weeks from three sets of 20sec with 1min rest for position one and three sets of 15sec with 30sec rest for position two in weeks one to three, progressing over the following weeks to a gradual increase in duration whereby in the seventh and eighth week the subjects were completing four sets of 1min with 1 min rest for position one and four sets of 30sec with 30sec rest for position two. The Hz remained the same throughout the eight-week period at 35Hz.

As outcome measures, three performance tests were performed initially and after eight weeks: counter-movement jump; extension strength of lower extremities with an isokinetic horizontal leg press; and a sit and reach test for flexibility. The results demonstrated that in the vibration group, whole-body vibration induced significant improvement of bilateral knee extensor strength, counter-movement jump and flexibility in the sit and reach test after the eight weeks of training. There was an improvement in flexibility measures of 13 per cent in the sit and reach test. No significant changes were found for any of the outcome measures for the control group.

So it appears that there is some good evidence relating to vibration training and its application for increased flexibility as well as performance measures. There is no doubt that this is an area that requires further research, with improved guidance relating to training and conditioning protocols, but for performance sport and recreational athletes who perhaps do not have the time to complete more traditional health-related fitness sessions, there may be some positive benefits in vibration training in relation to improved flexibility and also increased performance measures.

Vibration training stretch techniques: adductor stretch.

Squat on vibration platform.

Hamstring stretch on platform (single leg).

Hip flexor/quadriceps stretch on platform.

Hamstring stretch on platform (double leg).

Shoulder stretch on platform.

Calf stretch on platform.

Whole-Body Vibration Training Stretching Application

Rationale: increased ROM/flexibility in addition to improved performance measures.

Outcome: flexibility improvements/ increased ROM. Increased muscle activation to improve performance measures. And also studies imply improved health-related benefits, like improved body composition and bone density in post-menopausal women as well as strength gains/improvements. (More research is necessary to look at the duration of the gained benefits after treatment and also regarding specific performance and training benefits.)

Repetitions: one to two repetitions per muscle group/limb.

Hertz: range from 30–50Hz.

Amplitude: 2–4mm.

Duration: 20–60sec per limb/muscle group.

Rest: 15sec rest between each muscle group.

Frequency: 3 × week.

DOMS benefit: not relevant in vibration platform stretching.

PILATES AND YOGA

The rise in popularity of both Pilates and yoga has been significant over the past few years, with Pilates and yoga classes now being offered at leisure centres and sports clubs all over the world. Many people choose to take time out of their busy lives to focus on themselves for a 45 or 60min session to help with relaxation and well-being, and to have some 'me' time. The focus of many of these classes is mobility, strength, core stability and flexibility. But do these two disciplines offer effective techniques to assist in the development of flexibility training for the subjects involved?

DEFINITION OF PILATES

A system of exercises using special apparatus, designed to improve physical strength, flexibility and posture, and enhance mental awareness.

(*Oxford English Dictionary* [Oxford: OUP, 2015].)

PILATES

Joseph Pilates (1883–1967) defined physical fitness as 'the attainment and maintenance of a uniformly developed body with a sound mind'. His method has been around since the early 1900s. As a child, Joseph Pilates was a frail, sickly child prone to a number of illnesses including asthma, rheumatic fever and rickets. The doctors warned his parents that he had a short life expectancy. Rather than use the accepted norms of exercise to keep fit and healthy, he experimented with yoga, skiing, gymnastics, circus training, self-defence, weight training and dance. He developed his technique on the Isle of Man with soldiers injured in World War I and focused his work on improved health, rehabilitation and fitness. He took the opportunity of this time to develop his ideas, for example attaching springs to hospital beds that would enable bedridden patients to exercise against resistance. He also created strengthening exercises, which the doctors noticed improved the patients' speed of recovery. These experiments later formed the basis of the development of Pilates equipment.

The application in Pilates now is mainly for acute and chronic back pain, proprioception and postural benefits. Chartered Physiotherapists may become trained in Pilates and offer specific exercise regimes, either mat-based or on a reformer to help with postural control, strength, flexibility and stability.

Phrompaet et al.[77] assessed and compared the effects of Pilates exercise on flexibility and lumbo-pelvic movement control between the Pilates training and control groups. Forty subjects were randomly divided into Pilates-based training (twenty subjects) and the control groups (twenty subjects). The Pilates group attended 45min training sessions, two times per week for a period of eight weeks. Flexibility and lumbo-pelvic stability tests were determined as outcome measures, using a standard sit and reach test and pressure biofeedback respectively at zero, four and eight weeks of the study. As hamstrings tightness and low back flexibility are associated with low back pain, the introduction of Pilates seems to be a sensible conditioning option, if indeed the Pilates technique has a positive outcome on a subject's ROM and flexibility. The results showed that in the Pilates group, the measure score of the sit and reach test from baseline (week zero), week four and week eight were 27.69cm, 31.77cm and 34.89cm, respectively, an improvement of 7.2cm range from baseline to week eight. The Pilates group therefore improved flexibility significantly during this eight-week block. In the control group, the mean baseline (week zero) of sit and reach test was 22.74cm. The sit and reach tests at four and eight weeks of the study were 22.51cm and 22.91cm respectively, so no significant change in low back flexibility measures for the control group.

The Pilates method is a combination of static- and dynamic-stretching exercises which are safe to provide an increase in flexibility with subjects who may require a more controlled application of load and resistance. It can therefore be a beneficial training modality for both injured and non-injured subjects.

Pilates exercises focus on flowing movement throughout the whole body. The intensity of movement is the final range of motion at a tightness point without discomfort. The dosage for the study above by Phrompaet et al. was 45min training sessions, two times per week for a period of eight weeks, with five repetitions per position. They conclude that Pilates can be used as an adjunctive exercise programme to improve flexibility and enhance control mobility of trunk and pelvic segments. It may also prevent and attenuate the predisposition to axial musculoskeletal injury.

Kibar et al.[78] completed a randomized controlled study that aimed to determine the effect of Pilates mat exercises on dynamic and static balance, hamstring flexibility, abdominal muscle activity and endurance in healthy adults. They took twenty-four female subjects and prescribed 60min of mat-based Pilates two times each week for eight weeks and prescribed zero Pilates and purely normal daily living for twenty-three other female subjects. Measurements were recorded for dynamic and static balance, as well as the sit and reach test and curl-up test. A pressure biofeedback unit was used to measure transversus abdominis and lumbar muscle activity. In the comparison between the two groups, there were significant improvements in the Pilates group for sit and reach test and pressure biofeedback scores in week six, while in week eight there was also significant progress in the curl-up and static balance scores. Dynamic balance did not improve significantly.

Vaquero-Cristóbal et al.[79] performed a systematic review of Pilates practice and its effects, plus a detraining period, on hamstring extensibility, pelvic tilt and trunk flexion in maximal trunk flexion with knees extended. They analysed twenty-one different papers on the subject. It was found that the Pilates practice, with different volumes, significantly increased hamstring muscle extensibility and pelvic tilt in maximal trunk flexion. Three sessions a week for six weeks were necessary to obtain a high trunk inclination. Detraining took place when a decrease in hamstring extensibility and trunk flexion occurred from the second week of non-adherence. The review concluded that there is moderate evidence that Pilates is an effective method to increase hamstring extensibility, pelvic tilt and the degree of trunk flexion in maximal flexion positions in sedentary and recreationally active people and also to increase hamstring extensibility in athletes. As with any training modality, though, a period of rest or non-adherence to the programme will lead to detraining and a return to pre-activity levels.

Kloubec[80] wanted to determine the effects of Pilates exercise on abdominal endurance, hamstring flexibility, upper-body muscular endurance, posture and balance. He had fifty subjects who were recruited to participate in a twelve-week Pilates class, which met for 1hr two times per week. Subjects were randomly assigned either to the experimental (n = 25) or a non-Pilates control group (n = 25). Baseline measures were completed before the start of the study, which included the 1min YMCA sit-up test to assess abdominal endurance; leg lowering was selected as a marker to assess lower abdominal endurance and pelvic stability. Participants were also asked to perform the maximal number of push-ups they could do according to the American College of Sports Medicine assessment protocol. Women were allowed to do their push-ups from their knees, whereas men used the standard straight body form. To measure hamstring flexibility, the sit and reach test and supine hamstring extension test were used. Two methods of hamstring flexibility assessment were used, because there was uncertainty as to which method would be most sensitive. Postural analysis was determined by having all participants stand behind a clear postural analysis grid and be assessed at different angles with anatomical markers.

Subjects then performed the essential (basic) mat routine, consisting of approximately twenty-five separate exercises focusing on muscular endurance and flexibility of the abdomen, low back and hips during each class session. At the end of the twelve-week period there was a significant level of improvement in all variables except posture and balance. This study suggests that individuals can improve their muscular endurance and flexibility using relatively low-intensity Pilates exercises that do not require equipment or a high degree of skill and are easy to master and use within a personal fitness routine.

This is good news for all recreational athletes and otherwise sedentary individuals who are looking for positive health-related benefits from Pilates. It can also be applied to athletes, who would gain from improved muscular endurance and flexibility of the abdomen, low back and hips. There would undoubtedly also be a mental and physical wellness benefit as well, simply by the act of taking 45–60min of time for oneself a few times each week in the busy lives that we all lead.

Pilates Application

Rationale: body conditioning and muscular endurance, rehabilitation, postural and spinal issues/rehabilitation and flexibility/mobility.

Outcome: improved flexibility, abdominal endurance, hamstring flexibility, upper-body muscular endurance measures after regular attendance.

Repetitions: two to three sessions per week.

Duration: 45–60min classes.

Frequency: long-term benefit; research demonstrates benefits from eight to twelve weeks+ of application. Normal detraining one to two weeks after non-attendance.

DOMS benefit: not relevant in Pilates.

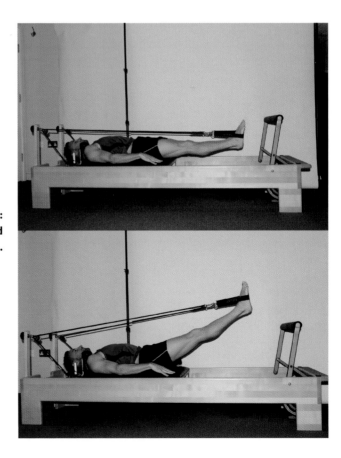

Reformer stretch exercises: long spine (straight legged version).

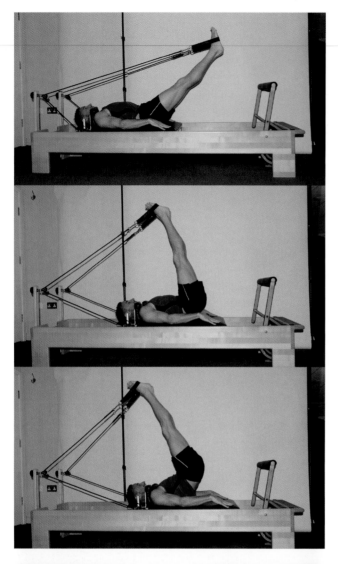

Long spine (straight legged version) (cont).

Static hip flexor.

Elephant stretch.

Russian split series.

Russian split series (cont).

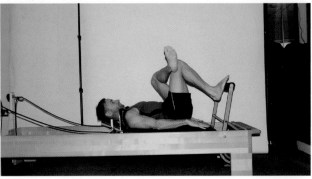

Glute stretch.

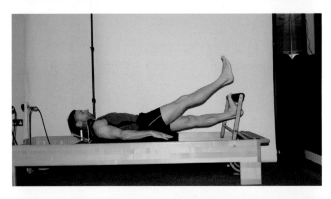

Calf stretch.

Pilates mat-based stretch exercises: arm-opening foot-lock progression.

Arm-opening foot-lock progression (cont).

Thread needle.

Thread needle (cont).

Saw.

Saw (cont).

YOGA

Yoga is an ancient form of exercise that focuses on strength, flexibility and breathing to boost physical and mental well-being. A series of movements designed to increase strength and flexibility form the main components of yoga, known as 'postures' in addition to breathing techniques. Yoga originated in India about 5,000 years ago and has been adapted in other countries in a variety of ways. There are many different styles of yoga, such as ashtanga, iyengar, hatha and sivananda, with some styles being more vigorous and physically demanding than others. Some may have a different area of emphasis, such as posture or breathing. Many yoga teachers develop their own practice by studying more than one style of yoga.

It is a very accessible form of exercise, with classes being offered in most leisure centres and sports facilities around the world. But does this exercise modality help with flexibility and mobility?

Much of the research completed with yoga appears to be focused on older populations, or those with chronic back pain. Farinatti et al.[81] compared the flexibility of elderly individuals before and after having practised hatha yoga and callisthenics for one year (fifty-two weeks), at least three times per week. Sixty-six subjects, of which twelve were men,

were measured and assigned to three groups: control (n = 24, age = 67.7±6.9 years); hatha yoga (n = 22, age = 61.2±4.8 years); and callisthenics (n = 20, age = 69.0±5.8 years). They were all assessed by the Flexitest.[82]

The maximal range of passive motion of thirteen movements in seven joints was assessed by the Flexitest, comparing the range obtained with standard charts representing each arc of movement on a discontinuous and non-dimensional scale from 0 to 4. (See 'Flexitest' in Chapter 3.)

Results of individual movements were summed to define four indexes (ankle and knee; hip and trunk; wrist and elbow; and shoulder) and total flexibility with the sum to create the Flexindex, as also discussed previously. The results indicated significant

---

DEFINITION OF YOGA

A Hindu spiritual and ascetic discipline, a part of which, including breath control, simple meditation and the adoption of specific bodily postures, is widely practised for health and relaxation.

(Oxford English Dictionary [Oxford: OUP, 2015].)

DEFINITION OF CALLISTHENICS

Gymnastic exercises to achieve bodily fitness and grace of movement.
(*Oxford English Dictionary* [Oxford: OUP, 2015].)

increases of total flexibility in the hatha yoga group (by 22.5 points on the normative scale) and the callisthenics group (by 5.8 points) and a decrease in the control group (by 2.1 points) after one year of intervention. Between group comparison showed that increases in the hatha yoga group were greater than in the callisthenics group for most flexibility indexes, particularly overall flexibility. Farinatti et al. conclude that the practice of hatha yoga (that is, slow/passive movements) is more effective in improving flexibility than callisthenics (that is, fast/dynamic movements and body weight movements), but callisthenics can also prevent flexibility losses observed in sedentary elderly subjects, as seen in the control group.

Tran et al.[83] looked at the effects of hatha yoga practice on the health-related aspects of physical fitness, including muscular strength and endurance, flexibility, cardiorespiratory fitness, body composition and pulmonary function. The study only used ten subjects, with an age range of between eighteen to twenty-seven years, but observed significant health-related fitness improvements from a minimum attendance of two yoga classes per week for a total of eight weeks, with the class duration being 1hr and 25min. Pre-class measurements were taken at baseline and reassessed after the eight-week block. All components of fitness improved significantly, with flexibility measures also improving markedly.

Flexibility was measured at the ankle, shoulder and hip joints. Three submaximal trials preceded three test trials for each movement. The average of the top two scores was used to represent maximal flexibility for that movement. Flexibility was assessed in a number of ways: a goniometer was used to measure the absolute range of motion at the ankle (plantar flexion and dorsiflexion). A measuring tape was used to measure shoulder elevation. In the trunk-extension test, the subject lay prone with feet and hips held down by researchers: with the hands clasped behind the head, the subject raised the trunk as far upward and backward as possible. Trunk extension was measured as the vertical distance between the mat and the chin. A sit and reach box was used to measure trunk flexion: while in the sitting position, with knees straight and in contact with the floor, the subject placed the index fingers of both hands together and reached as far forward as possible on the measuring tape, holding the position for 1 second. All flexibility measurements increased significantly, with ankle flexibility, shoulder elevation, trunk extension and trunk flexion increased by 13 per cent, 155 per cent, 188 per cent and 14 per cent, respectively. The results of this investigation indicate that eight weeks of hatha yoga practice can significantly improve multiple health-related aspects of physical fitness in young, healthy, predominantly female subjects. This is certainly a positive outcome for general health and well-being.

Gothe et al.[84] also looked at a comparison of the functional benefits of yoga compared to the conventional stretching–strengthening exercises recommended for adults. Sedentary healthy adults (N = 118; M age = 62.0) participated in an eight-week (three times a week for 1hr) randomized controlled trial, which consisted of a hatha yoga group (n = 61) and a stretching–strengthening exercise

Example of yoga pose: extended triangle pose.

group (n = 57). Standardized functional fitness tests assessing balance, strength, flexibility and mobility were administered at baseline and post-intervention. Once again, improvements were noted in all components of fitness. The conclusion is that regular yoga practice is just as effective as stretching–strengthening exercises in improving functional fitness. This is important for older and more sedentary populations as a yoga class may be a more desirable environment for this population, with both physical health-related benefits and perhaps a more social environment within a regular, structured yoga class. This may be helpful with class adherence and also in relation to general well-being within group-led exercise experiences.

Sherman et al.[85] looked at a comparison among yoga, stretching and a self-help book to aid in the recovery from chronic low back pain. A total of 228 adults with chronic low back pain were randomized to 12 weekly classes of yoga (92 patients), or conventional stretching exercises (91 patients), or a self-care book (45 patients). Back-related dysfunction measured via the Roland Disability Questionnaire (RDQ) declined over time in all groups. Compared with self-care, the yoga group reported superior function at twelve weeks and twenty-six weeks, while the stretching group reported superior function at six, twelve and twenty-six weeks. Sherman et al. conclude that physical activity involving stretching, regardless of whether it is achieved using yoga or more conventional exercises, has moderate benefits in individuals with moderately impairing low back pain. Yoga and stretching are therefore reasonable

Anusara Yoga

This form of yoga is known as the most spiritual forms of yoga than any other types. It focuses on your inner self, mind & soul

Kundalini Yoga

This yoga introduces you to your inner soul. There are about 7 types of yoga chakras included in Kundalini yoga. Following are the various poses included in this yoga

Vinyasa Yoga

The reason why Vinyasa yoga is different from all other forms of yoga is that it includes postures as well as breathing techniques

Bikram Yoga

This type of yoga is often practiced in a hot and humid environment where the temperature is about 40.6 degree Celsius

Hatha Yoga

The main purpose of this form of yoga is that it introduces beginners to yoga with the basic asanas & relaxation techniques

Five types of yoga and their benefits.

treatment options for persons who are willing to engage in physical activities to relieve moderately impairing back pain.

Grabara and Szopa[86] looked at the effects of hatha yoga exercises on spinal flexibility in women over fifty years old. The study included fifty-six women ranging in age between fifty to seventy-nine years of age and attending 90min hatha yoga sessions once a week. The range of spine mobility in three planes was measured before and after the intervention. This study showed that the applied yoga exercises increased spinal mobility and flexibility of the hamstring muscles regardless of age after a twenty-week intervention of just 90min once a week. They conclude that yoga exercises should be recommended to the elderly to make their muscles more flexible and to increase the range of motion in the joints, which is particularly important for improving quality of life. Their results indicated that practising yoga postures (asanas) even once a week led to an increase in the mobility of spinal joints and flexibility of the hamstring muscles. All subjects showed some improvement in flexibility regardless of age. This is the common theme regarding the benefits of yoga and is encouraging as an exercise modality of choice to assist in wellness and improved flexibility for injury prevention, injury management or quality of life in later years.

Yoga Application

Rationale: body conditioning, health-related fitness benefits, postural and spinal improvements and improvement in flexibility/mobility; benefitting all ages, especially sedentary and elderly individuals.

Outcome: improved flexibility, postural control and strength measures after regular attendance (>8 weeks).

Frequency: one to three sessions per week; even one session per week demonstrated improvements.

Class duration: >60min classes × 2/week or 90min classes × 1/week.

Long-term benefit: research demonstrates benefits from eight to twelve weeks+ of application and adaptation.

DOMS benefit: not relevant in yoga.

CHAPTER 5

INJURY PREVENTION AND PREPAREDNESS

An important area that I want to consider within this book is whether stretching could affect the outcome or prevalence of injury in adults and also in young people, during growth spurts for example. Specifically for youngsters, can we better prescribe age-appropriate stretching as a training modality to help with injury prevention? In addition, how could stretching perhaps help with injury prevention or injury management in adults?

For young people, sports participation both at school and with local clubs is very common and is certainly encouraged by parents for its health-related benefits, as well as its social and skill acquisition gains. The Centre for Disease Control and Prevention in America states that over 2.6 million children aged up to nineteen years old are treated in the emergency department each year for sports- and recreation-related injuries. Could these be prevented by improved flexibility as the child grows and develops? Indeed, how many of these incidents will be growth-related injuries and could we prevent them by effective stretching and flexibility programmes?

Adirim and Cheng[87] completed an overview of injuries in the young athlete. They discuss the fact that an ever-increasing number of children participate in sports and recreational activities, leading to a resultant increase in acute and overuse injuries. The most commonly injured areas of the body include the ankle and knee, followed by the hand, wrist, elbow, shin and calf, head, neck and clavicle. Contusions and strains are the most common injuries sustained by young athletes.

In early adolescence, apophysitis or strains at the apophyses are common. The most common sites are at the knee (Osgood-Schlatter disease [OSD]), at the heel (Sever's disease) and at the elbow (Little League elbow). Non-traumatic knee pain is one of the most common complaints in the young athlete.

DEFINITION OF APOPHYSITIS

An inflammation of an outgrowth, projection or swelling, especially a bony outgrowth that is still attached to the rest of the bone. Apophysitis occurs due to excessive traction or stress most frequently affecting the calcaneus (Sever's disease), the knee (Osgood-Schlatter), the shoulder (Little Leaguer shoulder) or elbow (Little Leaguer elbow).

(http://medical-dictionary. thefreedictionary.com/apophysitis)

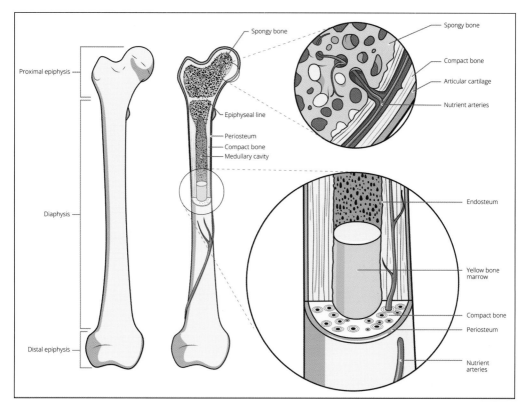

Bone structure.

Patellofemoral pain syndrome has a number of causes that include overuse, poor tracking of the patella, malalignment problems of the legs and foot problems, such as pes planus.

So can we prevent these overuse injuries that are so prevalent in young athletes? Nakase et al.[88] examined the precise risk factors for OSD by performing a prospective cohort study on a group of asymptomatic patients in particular times of adolescence using ultrasonography. Ehrenborg and Lagergren[89] described four radiological stages of tibial apophysis maturation:

1. cartilaginous stage
2. apophyseal stage
3. epiphyseal stage
4. bony stages.

These phases are normal developmental processes that occur as we grow and reach maturation. An ultrasound machine was used in the study to evaluate the current level of maturation for the subjects. A three-stage classification system for tibial tuberosity development was observed on ultrasonography: sonolucent (stage S); individual (stage I); and connective stages (stage C). 150 male soccer players (mean age 12.6±1.6 years, range 9–15 years) with 300 knees who practised soccer daily for 2hr were included in the study; 37 players (mean 11.2±1.1 years) with 70 knees at asymptomatic stage I were observed in the first examination. The result was 28 knees in stage S, 160 knees in stage C, 40 knees affected by OSD out of the 300 knees. Two knees were already showing

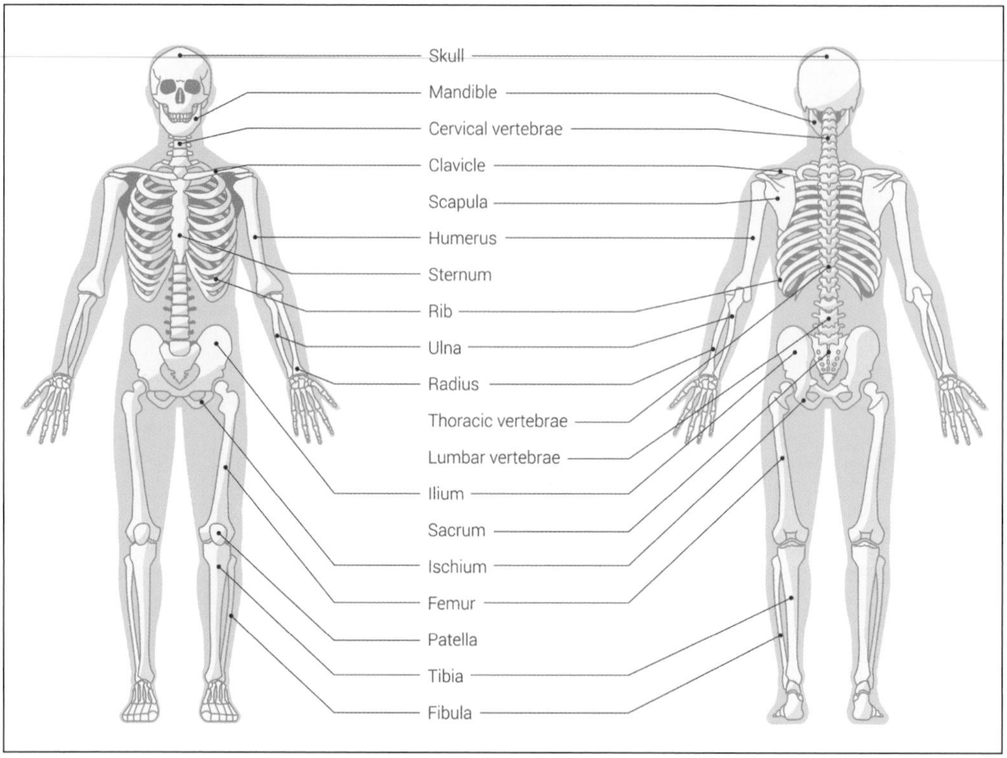
Skull
Mandible
Cervical vertebrae
Clavicle
Scapula
Humerus
Sternum
Rib
Ulna
Radius
Thoracic vertebrae
Lumbar vertebrae
Ilium
Sacrum
Ischium
Femur
Patella
Tibia
Fibula

Skeletal system.

completion of tibial tuberosity development on ultrasonography during the first examination, so they were excluded from the study.

The same subjects were re-evaluated via ultrasound a year later. A comparison of ten knees with OSD and sixty knees without OSD was completed. Height, body weight, body mass index, tightness of the quadriceps femoris muscle and hamstring muscles and muscle strength during knee extension and flexion were compared again. The incidence of OSD was 14.3 per cent in this one-year cohort study. Nakase *et al.* state that the most important finding was that increased quadriceps femoris muscle tightness, reduced muscle strength during knee extension and poor flexibility of the hamstring muscles are all risk factors for OSD. However, the researchers defined that the most important aspect for the prevention of OSD is an improvement of the flexibility of the quadriceps femoris muscle and conclude that an effective quadriceps stretching programme may help to prevent OSD in pre-adolescent footballers.

From reviewing the different stretching modalities, we may suggest that a myofascial release programme or vibration training programme may help in increasing ROM in the quadriceps muscles. Myofascial release is certainly a modality that is easy to prescribe and is an evidence-based practice for improvement of ROM.

Antich and Brewster[90] completed a review of the literature and physical therapy management for OSD. The literature once again focused on the muscular tightness of a growing adolescent, whose bones grow faster than their tendons can keep up with, as a possible causative factor. They discuss how physical therapy evaluation is outlined with pain-control techniques and how stretching exercises for the quadriceps and hamstrings muscles are prescribed for injury management. Antich and Brewster conclude that the disappearance of symptoms coinciding with the closure of the apophyseal plate is often the end result. The visual lump on the tibial tuberosity may still remain into adult life, but the associated pain will have subsided.

Posterior heel pain can be diagnosed as a number of things, from Sever's disease to Achilles tendinopathy amongst many others. Elengard et al.[91] discuss three main presentations of an injury: acute onset; chronic or overuse onset; and growth-related. They reviewed the treatment for posterior heel pain in young athletes and looked at a number of possible options, depending on many factors, like age, diagnosis and prognosis, then considered a number of possible treatment options, including stretching, strengthening, heel lifts and orthotics, activity modification and return to sport, as well as prevention of injury and re-injury.

In regard to stretching, the literature certainly encourages stretching of the calf musculature, stating that it is one of the main suggestions for treatment regardless of whether the symptoms come from the heel or from the Achilles tendon. If the subject has had a recent growth spurt prior to developing symptoms, the muscle-tendon unit might not have adjusted to the increased skeletal length, as mentioned above as a cause of growth-related injuries. This would then result in a tight calf musculature and Achilles tendon, along with a decreased ROM in ankle dorsiflexion, which may all lead to changed mechanics and possible resultant irritation and increased injury risk.

Some studies have shown that ROM in ankle dorsiflexion can be increased with stretching by approximately 1 degree, but the actual increase might just be due to increased stretch tolerance and compliance and/or lengthening of the muscle-tendon complex, which is what we have seen before with our review with most of the modalities of stretching, that is, a better capacity to cope with pain through an increased ROM. Do we increase ROM because we become more familiar with the sensation, or because we have increased the muscle length for the long term? Whether stretching in this patient population actually promotes increased range of motion and relieves symptoms needs to be further investigated. Elengard et al. recommend that a thorough evaluation needs to be performed for each subject and that stretching should be considered as a treatment only if there is a ROM deficit that needs to be addressed.

Park et al.[92] looked at the influence of stretching and warm-up on Achilles tendon material properties. They found that warming up the Achilles tendon before a number of objective measures were performed led to the same outcome measures, whether the individual warmed up with static stretching, or with a general, more dynamic warm-up. They hypothesized that warming up has a greater effect on Achilles tendon biomechanics than static stretching, but the results indicated that stretching, warm-up alone or combined, did not demonstrate statistically significant differences and may in fact have an equivalent effect on Achilles tendon biomechanics. They recognized that prolonged and more intense protocols may be required for changes to occur in each warm-up modality.

Previously, Park and Chou[93] had completed a review of the literature in regard to stretching for the prevention of Achilles tendon injuries. They discuss how few studies have addressed the effect of stretching in Achilles tendon injuries and that it is unclear if the conclusions made for musculoskeletal injuries can be applied to the Achilles tendon. Although many theories have been published regarding the potential benefits and limitations of stretching, few studies have been able to demonstrate definitively its utility in injury prevention. So are we stretching the muscles to help the tendons, or stretching the tendons themselves? And what will be the outcome?

Witvrouw et al.[94] looked at the role of stretching in tendon injuries. Tendons have an elastic component, which allows the stretch-shortening cycle to occur; this is an essential action in sports and activity, as previously mentioned. We need to create and build upon compliant tendons to help with performance and also to remain injury-free with the loads that are placed upon our tendons. We need a balance of 'stiffness' and elastic properties.

Witvrouw et al. place tendon properties and actions into two categories: tensile force transmission; and storage and release of elastic energy during locomotion. The action of tendons in storing and releasing energy is mainly seen in sports activities with stretch-shortening cycles. The more intense the movements of the stretch-shortening cycle (such as jumping-type activities), the more frequently tendon problems are observed. High stretch-shortening cycle movements impose high loads on tendons. Consequently, tendons that frequently deal with high stretch-shortening cycle motions require a high energy-absorbing capacity so as to store and release this large amount of elastic energy. Think of it like a spring or coil.

As the elasticity of tendon structures is a leading factor in the amount of stored energy, it means that prevention and rehabilitation programmes for tendon injuries should focus on increasing this tendon elasticity in athletes performing high stretch-shortening cycle

DEFINITION OF STRETCH-SHORTENING CYCLE

An eccentric muscle contraction followed immediately by a concentric contraction of the same muscle group. The elastic potentiation that occurs during the eccentric phase increases the force of output of the concentric contraction. These exercises replicate functional movement patterns and are typically used in the advance phase of rehabilitation, particularly in sports rehab. Exercises incorporating this phenomenon are called plyometrics.
(http://medical-dictionary.
thefreedictionary.com/
Stretch+shortening+cycle)

movements. The research review demonstrated a different tendon response to static stretching and ballistic stretching. The findings cited by Witvrouw et al. from work done by Mahieu et al.[95] provide evidence that static and ballistic stretching have different effects on passive-resistive torque and tendon stiffness. Eighty-one healthy subjects were randomized into three groups: a static-stretch group; a ballistic-stretch group; and a control group. Both stretching groups performed a six-week stretching programme for the calf muscles. Before and after this period, all subjects were evaluated for ankle ROM, passive-resistive torque of the plantar flexors and stiffness of the Achilles tendon.

The results of the study reveal that the dorsiflexion ROM was increased significantly in all groups. Static stretching resulted in a significant decrease of the passive-resistive torque; we know from our static-stretching review that power output is reduced as a

Stretch-shortening cycle concept: skipping rope is an example of a stretch-shortening cycle activity.

result of static stretching. However, with static stretching there was no change in Achilles tendon stiffness and it did not affect the outcome of this measure. In contrast, ballistic stretching had no significant effect on the passive-resistive torque of the plantar flexors, so there was no loss of power, again confirming what we already know from ballistic stretching. However, a significant decrease in stiffness of the Achilles tendon was observed in the ballistic-stretch group, allowing the tendon to become more elastic, which is a desirable outcome. Consequently, both types of stretching could be considered as complementary for rehabilitation programmes when the aim is an increase in ROM, but ballistic stretching allows for more positive performance gains. So we can increase ROM by both static and ballistic applications, and improve the tendon energy-absorbing capacity via ballistic stretching.

So if we utilize an appropriate stretching modality as part of our preparation, we can perhaps affect tendon capacity and reduce the incidence of injury. Additional techniques to assist in tendon protection would include speed, agility and quickness conditioning and plyometric training, which have dynamic properties similar to ballistic stretching, so as to encourage the elastic properties described above and also to develop the desired increase in tensile force production.

Example movements for speed, agility and quickness to develop the stretch-shortening cycle: speed ladder.

Bounding.

Hurdle jumps.

Hurdle double leg jump.

The stiffer the tendon, the faster the force is transferred to the bones and the more efficient the concentric contraction becomes to allow explosive movement. In this way, the metabolic energy of the muscle is converted efficiently into mechanical work, which is a desirable commodity in sports performance and may also positively affect injury-prevention strategies. Indeed, I prescribe speed, agility and quickness or skipping, for example, to all of my lower-limb tendon clients to help improve the tendon capacity. Appropriate ballistic or dynamic stretching may also positively affect this capacity.

Kamonseki et al.[96] looked at the effect of stretching with and without muscle strengthening of the foot alone, or foot and hip, on pain and function in patients with plantar fasciitis. Stretching the calf muscles is often prescribed to help improve plantar fasciitis, so it is interesting to look at the outcome from this recent study. Eighty-three patients with plantar fasciitis were allocated to one of three treatment options for an eight-week period: foot exercise group (extrinsic and intrinsic foot muscles); foot and hip exercise group (abductor and lateral rotator muscles); and a stretching alone exercise group. Acute measures were taken, with a visual analogue scale for pain, the Foot and Ankle Outcome Score and the Star Excursion Balance Test. All evaluations were performed before treatment and after the last treatment session. In this study, all three exercise protocols that were analysed led to improvements at the eight-week follow-up in pain, function and dynamic lower-limb stability in patients with plantar fasciitis. So calf stretches should certainly be included in an injury-management plan for plantar fasciitis.

Celik et al.[97] also reviewed how plantar fasciitis responds to stretching as a treatment modality. They looked at a comparison of the effectiveness of joint mobilization combined with stretching exercises versus steroid injection in the treatment of plantar fasciitis. A total of 43 patients (mean age, 45.5 ± 8.5 years; range, 30–60 years; 23 females) with plantar fasciitis were randomly assigned to receive either joint mobilization combined with stretching exercises (n = 22) or a steroid injection (n = 21). Joint mobilization combined with stretching was applied three times per week for three weeks for a total of nine visits and the steroid-injection group received one injection at baseline only. The patients' functional scores were assessed using the Foot and Ankle Ability Measure and pain was evaluated using the Visual Analogue Scale. Outcomes of interest were captured at baseline and at week three, week six, week twelve and one-year follow-ups.

The results of the study demonstrated significant improvements in pain relief and functional outcomes in both groups at the three-, six- and twelve-week follow-ups compared to baseline. However, at the twelve-week and one-year follow-ups, pain and functional outcomes were significantly improved in only the joint mobilization combined with stretching group. This study therefore demonstrated that while both groups achieved significant improvements at the three-, six- and twelve-week follow-ups, the noted improvements continued in only the joint mobilization combined with stretching group for a period of time ranging from twelve weeks to one year. This implies that there will be a more successful outcome over time in treating plantar fasciitis by using a joint mobilization and stretching injury-management plan, rather than a steroid injection. More research is certainly required in this area.

Behm et al.[98] completed a systematic review of the acute effects of muscle stretching on physical performance, ROM and injury incidence in healthy active individuals. They found twelve studies which looked at the influence of pre-activity stretching on injury risk and recognized that stretching is generally incorporated into the pre-activity routine in numerous sports. Eight out of the twelve studies showed some effectiveness of stretching pre-activity, whereas four showed no effect. Of practical importance, there was no evidence in the studies that stretching would negatively influence injury risk.

There were, of course, limitations in the review of the different studies as a result of varied research protocols, stretch durations and stretching with or without additional warm-up, plus many of the studies reviewed different types of injuries. Of the eight studies examining the effect of stretching on total injury rates, only two reported a benefit of stretching. Six studies specified the effects of stretching on the prevalence of acute muscle injuries. From these studies, it was possible for the review panel to compute the relative risk of sustaining an acute muscle injury associated with stretching versus not stretching. Taken together, these studies indicate a 54 per cent risk reduction in acute muscle injuries associated with stretching. This is a significant risk reduction.

Because of the variable studies, it is difficult to determine if this is because of stretching alone, or combined with warm-up and therefore improved viscoelastic properties, as discussed previously within the review of dynamic stretching. When reviewing a comparison in the literature between endurance activities with a predominance of overuse injuries versus sprinting sports with a high prevalence of muscle injuries, the current research indicates that pre-activity stretching may be beneficial for injury prevention in sports with a sprint running component, but not in endurance-based running activities, which have a predominance of overuse injuries.

We have discussed above that there are a number of manageable and controllable components that may affect the outcome of an overuse injury, these being an inclusion of stretching, strengthening, appropriate orthotic prescription and activity modification, to name a few. Certainly the occurrence of an acute injury may be reduced by pre-activity stretching, with appropriate ROM activities occurring in the individual who is participating in the sport/activity. The benefit of warm and conditioned muscles would allow for injury-free activity. It is when a poorly conditioned individual attempts a movement beyond their capacity at end ROM, for example, that acute injuries may occur.

So when injury is prevalent in sporting activities, of course we want to prevent these as much as possible. In professional sport, an injured player is generally a loss to the team, but also costs money. It is very important for recreational athletes and sports teams to consider injury-prevention strategies as part of their warm-up and different stretching modalities within the warm-up may be an important component of this.

If we look at the governing body for football, FIFA, they have developed through the FIFA Medical and Research Centre (F-MARC) a warm-up protocol called the FIFA 11+. It is a complete warm-up programme targeted at reducing injuries among male and female football players aged fourteen years and older. The programme was developed by an international group of experts and should be performed, as a standard warm-up, at the start of each training session at least twice a week.

The warm-up takes around 20min to complete. But does it help to reduce the incidence of injury for youth footballers? Barengo et al.[99] completed a systematic review of the impact of the FIFA 11+ training programme on injury prevention in football players. The aim of this systematic review was to evaluate the impact of the FIFA 11+ on injury incidence, compliance and cost-effectiveness when implemented among football players. They found in their review that the FIFA 11+ has effectively demonstrated how a simple exercise programme completed as part of warm-up can decrease the incidence of injuries in amateur football players. In general, considerable reductions in the number of injured players, ranging between 30 per cent and 70 per cent, have been observed among the teams that implemented the FIFA 11+. In addition, players with high compliance to the FIFA 11+ programme have an estimated risk reduction of all injuries by 35 per cent and demonstrate significant improvements in components of neuromuscular and motor performance when participating in structured warm-up sessions at least 1.5 times/week.

Interestingly, the FIFA 11+ has only two dynamic-stretching exercises within the entire 20min warm-up routine: hip out (hip abduction); and hip in (hip adduction). No other mode of stretching was prescribed or performed by the players within the protocol. Other components of the warm-up include: running movements; shoulder contact; strength (core); plyometric and balance work; specific hamstring strengthening; proprioception; lunges and squats; single leg and double leg jumps; higher intensity running; bounding; and cutting. The warm-up course is over the width of a football pitch and the players complete each exercise in pairs for a total duration of 20min. Is the benefit of the programme that the players are becoming better conditioned for the sports-specific movement patterns that football presents, which in turn reduces the incidence of injury? And are they more appropriately warmed up and prepared for training?

The injury-prevention data with the FIFA11+ is encouraging, especially in this age group, who are perhaps more susceptible to injuries because of growth, physical development and differing levels of maturation. There are also the challenges of possible high training volume and load, especially for the more talented players, who typically get asked to play and train more often. Indeed, the more talented an individual is, the more often they tend to get asked to perform in their teams or in different sports, so their injury risk may therefore be increased by the physical demands placed upon them. More information and a detailed description of the 11+ protocol and appropriate application can be found at http://f-marc.com/11plus/11plus/.

Lauersen et al.[100] completed a systematic review and meta-analysis of a randomized controlled trial looking at the effectiveness of exercise interventions to prevent sports injuries. They wanted to determine whether physical activity exercises could reduce sports injuries, using stratified analyses of strength training, stretching, proprioception and combinations of these, to provide separate acute and overuse injury estimates. It was deemed that 25 trials would be appropriate for comparison; these included 26,610 participants with 3,464 injuries. From their review, stratified exposure analyses proved to have no beneficial effect for stretching, whereas studies with multiple exposures, proprioception training and strength training showed a tendency towards increasing effect. Both acute injuries and overuse injuries could be reduced by physical activity programmes.

In these evidence-based reviews, despite a few outlying studies, Lauersen et al. established that consistently favourable estimates were obtained for all injury-prevention measures except for stretching. The data does not support the use of stretching for injury-prevention purposes, neither before nor after exercise. However, it should be noted that this analysis only included two

Summary of Injury Management			
Target Area for Injury Prevention	**Requirement for Injury Prevention/ Management**	**Stretching Modality of Choice**	**Outcome**
Chronic/Overuse Injury	Appropriate/ functional ROM and appropriately conditioned individual.	PNF Myofascial release Vibration training.	Improved ROM to increase muscle length and release associated tissues.
Acute Injury	Appropriately conditioned individual.	Dynamic/ballistic stretching pre-exercise with an aim of increased body temperature and readiness. Vibration platform training stretching, PNF or myofascial release may be completed before exercise if an increased ROM is desired.	No indication for injury prevention. Increased ROM and physical preparedness for activity. Research still needed within this area.
Growth/ Age-Related Injury	Maintain tissue extensibility as growth occurs to prevent growth injury. Appropriately conditioned individual.	PNF Myofascial release Vibration training (all to increase ROM with no compromise on performance).	Reduced incidence of age-related growth injuries.

studies on army recruits and one internet-based study on the general population because of study selection criteria and too much variation in the research for comparison, which is always a challenge in these reviews. Strength training reduced acute sports injuries to less than one-third and overuse injuries could be almost halved based upon their review.

So it seems that with acute injury-prevention stretching is deemed to be irrelevant, which I will discuss further below in the 'Management of Acute Injuries and Stretch-ing'. Appropriate sports-specific warm-ups seem much more important.

For growth-related and chronic or overuse injuries, stretching seems to have a purpose and a relevant, positive outcome to assist with increased ROM to relieve the pressure on the tendons of the growing athlete and help to prevent growth injuries. Increased ROM, achieved with stretching, is beneficial for injury management for chronic or overuse injuries, so a relevant stretching modality should certainly be prescribed in this youth age group range.

MANAGEMENT OF ACUTE INJURIES AND STRETCHING

One of the most confusing areas that individuals appear to need guidance with is when to stretch after experiencing an acute injury. So, for example, if someone experiences a grade I hamstring strain, what should they do? I hear clients and individuals all the time comment 'I am stretching it already', as if this is the golden nugget to their recovery and rehabilitation. But, is it? When should you stretch with an acute injury and does the grade of injury also affect what to do?

Treatment should address all soft tissue injuries immediately with the same principle, that of PRICE:

- P: Protection
- R: Rest
- I: Ice
- C: Compression
- E: Elevation.

The goal is to stop the bleeding within the muscle as soon as possible and to minimize any damage within the tissues. It is important to consider that there are three main phases in the repair of a torn muscle: the inflammatory response phase; the proliferative/regeneration phase; and the remodelling phase. If and when to stretch is determined by taking into account the phase that the muscle repair is in. We certainly do not want to be counterproductive when managing a new injury, by re-tearing healing tissue and prolonging recovery.

Järvinen et al.[101] discuss the importance of regeneration of skeletal muscle after injury. They explore how the recognition of some basic principles of skeletal muscle regeneration and healing processes can considerably

help in either avoiding re-injury or prolonged injury, and how this will affect the time frame before a return to competition. They discuss the importance of understanding the biology and physiology of muscle regeneration, in the hope of extending these findings to clinical practice in an attempt to propose an evidence-based approach for the diagnosis and optimal treatment of skeletal muscle injuries. Many practitioners will use their clinical experience and judgement to decide when it is appropriate to graduate a muscular injury from immobilization to progressive function. Järvinen et al. discuss the process of the healing pattern in their research:

1. **Destruction phase (also known as inflammatory response phase)**. The ruptured myofiber becomes necrotized only over a short distance. The propagation of the necrosis is halted by a 'fire door', a contraction band formed within a couple of hours, in the shelter of which the rupture is sealed by a new sarcolemma. The ruptured myofibers contract and the gap between the stumps is filled by a haematoma. The injury induces a brisk inflammatory cell reaction.

2. **Repair phase (also known as proliferative/regeneration phase)**. This begins with phagocytosis of the necrotized tissue by blood-derived monocytes. The myogenic reserve cells, satellite cells, are activated and begin the repair of the breached myofiber. Firstly, so-called committed satellite cells begin to differentiate into myoblasts. Secondly, undifferentiated stem satellite cells begin to proliferate by 24 hours and thereafter contribute to the formation of myoblasts, at the same time providing new satellite cells by asymmetric cell division for future needs of regeneration. The myoblasts arising from the committed and stem satellite cells then

DEFINITION OF MUSCLE STRAIN GRADES

Muscular strains are categorized into three different grades depending upon the severity of the damage to the muscle fibres.

Grade I Muscle Strain

There is damage to individual muscle fibres (less than 5 per cent of fibres). This is a mild strain that requires two to three weeks' rest.

Grade II Muscle Strain

There is more extensive damage, with more muscle fibres involved, but the muscle is not completely ruptured. The rest period required is usually between three to six weeks.

Grade III Muscle Strain

This is a complete rupture of a muscle. In a sports person this will usually require surgery to repair the muscle. The rehabilitation time is around three months.
(www.physioroom.com)

fuse to form myotubes within a couple of days. Within 5–6 days the necrotized part of the ruptured myofiber inside the remaining old basal lamina is replaced by the regenerating myofiber, which then begins to penetrate into the connective tissue scar between the stumps of the ruptured myofibers. The injury site is also re-vascularized by ingrowing capillaries with the first angiogenic capillary sprouts seen three days after the injury.

3. **Remodelling phase**. This is the period of maturation of the regenerating myofibers, which includes formation of a mature contractile apparatus and attachment of the ends of the regenerated myofibers to the intervening scar by newly formed myotendinous junctions. The retraction of the scar pulls the ends closer to each other, but they appear to stay separated by a thin layer of connective tissue to which the ends remain attached by newly formed myotendinous junctions. The contraction of the large granulation to scar tissue is driven by fibroblasts transforming to the contraction capable myofibroblasts.

It is important to allow these phases to occur so as to ensure that a strong and stable repair is progressed within the muscle. If the process is delayed, this can affect the outcome or duration of the injury. If the regenerating microfibre is not allowed to form the appropriate strength for a reintroduction of contraction, then re-rupture of the muscle fibres may occur.

Despite this, a reintroduction to activity and mobilization of the injury site should be completed gradually and within the limits of pain as soon as possible. Early mobilization has been shown best to expedite and intensify the regeneration phase of the injured skeletal muscle fibres, as well as to induce increased blood supply to the injured area. This decision by the practitioner regarding when to begin to mobilize the injured tissue is largely based upon clinical experience, as well as clinical presentation and the history of the individual. Yet this decision is crucial to the positive outcome for the injured individual.

Excellent body awareness on the part of the injured person is essential. They need to recognize when it is okay to progress and when they should limit their activity. During my experiences in rehabilitation, if the injured individual understands how their body feels and appreciates the difference between pain and muscular overload, then generally they will have a better rehabilitative outcome. Hamstring and calf injuries are great examples of this. I run an accelerated hamstring rehabilitation protocol for my injured athletes, in which I will get them moving by day one or two if cleared clinically by the chartered physiotherapist or athletic trainer. As discussed, this has to be pain-free, but they will walk or complete a very slow jog *pain-free*! This helps with circulation, as mentioned above, but also as long as this mobilization and controlled movement is pain-free, it will help to encourage a strong bond within the remodelling tissues. It is also beneficial psychologically for the athlete to get outside and start moving again, as usually the psychological burden of an injury on an athlete often does not help the healing process. By giving them appropriate, gradual activity to complete, even a pain-free prescribed duration or volume walk outside will have a positive effect.

Indeed, Sherry and Best[102] compared the effectiveness of two different rehabilitation programmes for acute hamstring strain by evaluating time needed to return to sports and re-injury rate during the first two weeks and the first year after return to sport. They used twenty-four athletes with an acute hamstring strain and randomly assigned them to one of two rehabilitation groups. Eleven athletes were assigned to a protocol consisting of static stretching, isolated progressive hamstring resistance exercise and icing, while thirteen athletes were assigned to a programme consisting of progressive agility and trunk-stabilization exercises and icing. The number of days for full return to sports, injury recurrence within the first two weeks, injury recurrence within the first year of return-

ing to sports and lower-extremity functional evaluations were collected for all subjects and compared among groups.

The results were very interesting and should certainly be considered for future applied rehabilitation protocols in relation to hamstring injuries, which are very common in both professional sports and amateurs alike. Sherry and Best found that the average (+/− SD) time required to return to sports for athletes in the static stretching, isolated progressive hamstring resistance exercise and icing group was 37.4 +/− 27.6 days, while the average time for athletes in the progressive agility and trunk-stabilization exercises and icing group was 22.2 +/− 8.3 days. However, this difference was not statistically significant.

In the first two weeks after return to sports, the re-injury rate was significantly greater in the static stretching, isolated progressive hamstring resistance exercise and the icing group, where six of eleven athletes (54.5 per cent) suffered a recurrent hamstring strain after completing the stretching and strengthening programme, as compared to none of the thirteen athletes (0 per cent) in the progressive agility and trunk-stabilization exercises and icing group. After one year of return to sports, re-injury rate was significantly greater in the static stretching, isolated progressive hamstring resistance exercise and icing group. Seven of ten athletes (70 per cent) who completed the hamstring stretching and strengthening programme, as compared to only one of the thirteen athletes (7.7 per cent) who completed the progressive agility and trunk-stabilization programme, suffered a recurrent hamstring strain during that one-year period. So, effective and appropriate conditioning appears to lead to a better outcome than stretching and specific hamstring resistance exercises in a return to sports post-hamstring injury.

The same concept from Sherry and Best was continued in 2013 by Silder et al.[103] (with Sherry included in the authors), where they looked at the original but slightly modified progressive agility and trunk-stabilization exercises from Sherry's original research and compared it this time to a progressive running and eccentric strengthening programme, with the goal being to determine which protocol promoted muscle and functional recovery more and minimized re-injury risk, while also optimizing athletic performance. The two rehabilitation programmes employed in this study yielded similar results with respect to hamstring muscle recovery and function at the time of return to sport. What was discovered from magnetic resonance imaging post-rehabilitation was that there was evidence of continued muscular healing after completion of the rehabilitation, despite the appearance of normal physical strength and function on clinical examination. This is important to consider when an individual returns to activity.

For me, an individual's rehabilitation and performance measures, within their training and conditioning, must continue to be addressed and improved on a regular basis even when they return to full activity. A return to training does not mean that rehabilitation is complete. Re-injury was also low for both rehabilitation groups after a return to sport.

Stretching was prescribed before the running in the progressive running and eccentric strengthening group. It included 5min of gentle stretching before and after each running session 3 × 20sec each for the following stretches:

- standing calf stretch
- standing quadriceps stretch
- half-kneeling hip flexor stretch
- groin or adductor stretch
- standing hamstring stretch.

Stretching was included in the progressive agility and trunk-stabilization group via more functional movement patterns, for example with a lunge walk that required trunk rotation and pelvic control with the hamstrings in a lengthened position, or single leg windmill touches. A return to functional ROM is important and the rationale for inclusion of the five static stretches for 20sec before running is not clear, but presumably it was to return normal range to the injured limb. However, both rehabilitation protocols had a very similar outcome in regard to minimal re-injury risk and optimal athletic performance post-injury.

Malliaropoulos et al.[104] looked at the effects of stretching in the rehabilitation of hamstring injuries. A grade II strain of the hamstring muscles allowed for inclusion within the study. A total of fifty-two male and twenty-eight female athletes were examined within the first 48hr period post-injury. Diagnosis was based upon injury mechanism: indirect muscle strain injury (usually after an eccentric contraction); clinical evaluation; the measurement of the active knee extension using goniometry in comparison with the uninjured limb; and the dimensions of the rupture in the ultrasound scan of the muscle. Reduction in the knee extension up to 10–15 degrees and echographic rupture dimensions up to 2.5cm width and 3.5cm length were diagnostic for a second-degree hamstring strain.

During the first 48hr, all of the athletes underwent the PRICE protocol as mentioned above, followed by a re-evaluation of the injury. The athletes were randomly assigned into two groups, A (n = 40) and B (B = 40) and followed the same rehabilitation programme. They carried out static stretches of the hamstring muscles, sustained for 30sec, repeated four times. The only difference between the two groups was the number of daily stretching sessions. Group A had four stretching session, whereas group B had four sessions daily. Static hamstring stretching was performed in a standing position in an anterior pelvic tilt, with the stretching leg on a chair or a table, depending on the athlete's height. The athletes were advised to stretch until they felt tension or slight pulling, but no pain. Two parameters were measured and studied in order to assess the effectiveness of the stretching programmes in the two groups: a) time needed for equalization of active knee extension between injured and healthy sides; and b) time required for full rehabilitation.

The results were established as follows: group A patients returned to normal values of ROM in a mean time of 7.3 days, whereas the respective time for group B was 5.6 days. Group B completed the stretches four times daily compared to once daily with group A. This difference between the two groups was statistically significant. The time required for full, unrestricted athletic activities was also significantly different between the two groups. Group B was again able to achieve a more prompt return to unrestricted activities, with the return to activity being 15 days for group A and 13.3 days for group B. So ROM gain post-injury when a muscle may be inhibited because of pain and loss of function appears to be an important component to a successful return to unrestricted athletic activities. Interestingly, once normal ROM was achieved, it still took each group the same amount of days to return to full unrestricted athletic activity, but group B reached this measure sooner because of the more prompt gain in active knee extension. Both groups took 7.7 days to return to full unrestricted activity once full ROM was achieved.

Malliaropoulos et al. have not indicated at what point post-injury (days) they started the stretching protocol within their studies, only that it was completed at a point when the participants could perform within a pain-free ROM with tension placed through the muscle during the stretch. They mention the second phase of recovery with a muscle tear, that being the repair/regeneration phase, and how stretching and appropriate pain-free stress is important to assist in strengthening and realigning the collagen fibres within the tissue for a successful rehabilitation outcome. They also discuss the stretch mode of choice and acknowledge that PNF may be a more appropriate stretching mode to regain ROM, but they felt that there was more control with static stretching post-injury.

It is also important to consider that this is one study of eighty people, so much more research is needed to become fully informed in best practice, but it certainly makes sense that a return to pre-injury ROM would certainly be desirable post-muscular injury and may accelerate a safe return to sport. Malliaropoulos et al. also do not follow-up on re-injury rate from their cohort, which is also a very important component to consider in an applied setting when our goal is to return our athletes back to action in a safe and prompt way, but with minimal risk of re-injury.

Valle et al.[105] wanted to propose a rehabilitation protocol for hamstring muscle injuries based on current basic science and research knowledge regarding injury demographics and management options. They felt that hamstring injuries were poorly managed and they wanted to accumulate all relevant information in order to create a criteria-based (subjective and objective) progression through the rehabilitation programme from initial injury to return to play. A pain-free ROM and strength gains are included as relevant objective measures; these are important criteria towards a safe return to play, so should be considered and prescribed to athletes in applied settings.

For a calf strain, Nsitem[106] discusses a case study of a male aged forty-four years, who presented with acute calf pain with a palpable defect, loss of range of motion and loss of strength after sustaining a soft tissue injury to the lower leg. The patient was treated over a six-week period. Initially, rehabilitation was approached using the PRICE principles for symptomatic relief, followed by stretching, strengthening, proprioception and conditioning exercises. In week one, the patient began active ROM exercises for the knee and ankle in the pain-free range. This pain-free active mobilization was completed before any isometric, concentric or eccentric strengthening exercises were introduced. These were introduced in weeks three to four. ROM progressions were included thereafter as function and strength improved. In weeks five to six, the patient achieved pain-free full ROM in the affected leg, full function and a return to training.

This rehabilitation protocol mirrors the same concept for hamstring tears, with a focus on progressive pain-free ROM to aid recovery and a return to function. This, in combination with specific strengthening exercises and a return to functional activity, is essential in rehabilitation and return to play plans.

Acute Injury and Stretching Application

Mode: static-active stretching, progressing to static-passive stretching relative to injury site, progressing to more functional movements, isometric, concentric and eccentric training, plus sport-specific ROM demands.

Rationale: to improve post-injury ROM, which will be reduced as a result of swelling and bleeding around the injury site, so this normal functional range needs to be regained.

Outcome: to improve post-injury ROM and return to pre-injury functional range. The subsequent realignment of collagen and graduated load and stress applied from stretching and load to the injury site will lead to a more positive outcome during rehabilitation, as long as appropriate strength progressions and load are placed upon the injury site. Consideration of pain-free movement and range, and progressive tolerance to load can help success.

Sets, repetitions and frequency: as clinically appropriate: pain-free ROM. 4 x 30 sec daily.

Long-term benefit: to return to pre-injury ROM and full pre-injury function with reduced risk of re-injury.

STRENGTH TRAINING TO IMPROVE RANGE OF MOVEMENT

When I have a new client who may have extremely stiff hamstrings or tight hip flexors, I often use strength exercises to help them increase their ROM, after consultation with their chartered physiotherapist, athletic trainer or doctor. This is achieved in a functional capacity, by completing the desired movement with progressive appropriate load, which does not compromise technique, but may increase joint ROM over time.

This concept can be addressed with almost any part of the body. Shoulders, thoracic spine and hamstrings, as well as hip flexors, tend to have reduced ROM for a number of reasons: muscle imbalance; previous injury; and dominant 'beach muscle' training rather than a balanced training regime, for example. Appropriately prescribed strength exercises can help to improve this limited ROM. This would be classed as static-passive stretching, where the barbell or dumb-bells would act as the load to increase the ROM.

We know that it is a challenge to achieve long-term changes in muscle length and PNF; myofascial release and vibration training are possibly the most effective stretching modes to achieve a change in ROM, so can strength training and specifically eccentric strength training improve flexibility?

O'Sullivan et al.[107] examined the evidence that eccentric training has demonstrated effectiveness as a means of improving lower-limb flexibility. Studies evaluating flexibility using both joint ROM and muscle fascicle length were included in their review. Six studies met the inclusion/exclusion criteria. They found that there was consistent, strong evidence from all six trials in three different muscle groups that eccentric training can improve lower-limb flexibility. They assessed the outcome using either joint ROM or muscle fascicle length. This is a positive step in conditioning and should be considered in exercise prescription for any individuals demonstrating reduced ROM.

2. Mid-position.

3. Full ROM as comfort allows.

1. Start position.

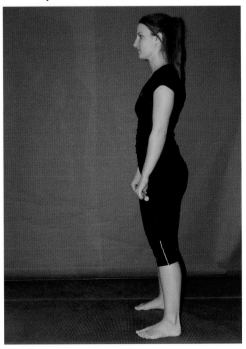

Straight leg dead lift: eccentric exercise that can help improve hamstring flexibility.

CASE STUDY

We will look at the case of a young man, an ex-professional rugby player with low back pain who was transitioning from rugby to more running-based activities. He presented as very lean and muscular. His injury history was of nil significance regarding his low back pain, meaning no acute injury or history of injury. However, he presented with a significant posterior pelvic tilt and lordosis, and he suffered from backache after long extended runs, when he would experience extended load through his increased pelvic tilt/lordosis for a duration >15min. After a physio review, a conditioning plan was put together for him, which focused on long-lever hip flexor exercises – Bulgarian squat and split-squat movements in particular.

The athlete was instructed to complete these exercises with a neutral spine, with a broomstick over his shoulders to ensure excellent posture and control, and what I call 'good branding' (chest out!). His rear leg extension position needed to be elongated and exaggerated, with a focus on anterior pelvic tilt to use the strength exercise to help increase his hip flexor ROM and mobility. After three weeks of completing his prescribed strength and conditioning programme, and specific strength-mobility session 3 × each week, 3 × 12 repetitions left and right on each exercise, with a focus on full ROM rather than load, he experienced a reduction in lumbar pain and increased hip extension mobility and is now pain-free. His Thomas test was improved, as well as the pain having gone doing running-based activities.

Elongated Bulgarian squat to help with improved hip flexor ROM.

We are aware that eccentric training will illicit DOMS, so this needs to be considered when prescribing eccentric exercises and explained to the individual that this is a normal response to this kind of load.

So what about athletes? How does a gymnast become as flexible as they are? Is this as a result of years of training at a young age when form follows function in the body, so as a young athlete it is easier to adapt to the ROM requirements and demands within a particular sport? I would like to provide a personal example. I took my daughter, aged six, to a local gymnastics club. I saw young gymnasts, maybe eight to nine years old, being forced into uncomfortable ranges of movement, with their shoulders and groins in a split position, and being pushed to a point of pain by their coaches. Is this acceptable to the growing athlete? Will their stretch tolerance

2. Mid-position.

1. Start position.

3. Position of tightness slight forward lean as a result of tight thoracic spine.

Overhead squat for improved thoracic mobility.

Gymnast in stretch position.

and subsequent ROM improve? Or is the physical development of these young athletes being put at risk? Is this how they become as flexible as they need to be to reach the top of their sport? As a strength and conditioning coach, what I saw shocked me. I did not want my daughter to be pushed to a point of pain when doing gymnastics. I wanted her to have fun, but maybe that's what defines the difference between the recreational athlete and the highly trained athlete?

British Gymnastics has a Long Term Athlete Development (LATD) model, which provides age-appropriate guidelines for parents, participants and coaches. Gymnastics is an early-specialization sport, so the physical demands may be more than in other sports at a young age. However, the British Gymnastics LTAD model sets out the fundamental physical pathway for anyone involved in gymnastics.

Flexibility is noted as an important component of the LTAD model from age seven onwards in the learning to train phase. The goal is to acquire the basic skills of gymnastics, based upon an age range of between seven to nine years. The training to train phase, which is aimed at ages ten to thirteen, in which the goal is to acquire key gymnastic skills, also has flexibility as an area to refine and address. It is noted within the pathway to consider the level of maturation of the participant and to take note of growth spurts, discipline, level of maturity and aspiration within each phase for each participant. From fourteen years onwards, the LTAD pathway does not specify flexibility as an individual component of fitness to be addressed, only for individual programmes to be considered for each participant. This age-group range is considered to be an acquiring advanced skills phase.

Baseball pitcher demonstrating a functional yet extreme ROM of the shoulder joint.

There is no doubt that by training and performing specific movements in an increased, functional ROM, the body will adapt and this functional, sometimes extreme ROM, will become the norm. For example, if I tried to do the splits, well, I just couldn't! But for a conditioned gymnast it would be easy and is what they do. But for ballet dancers, footballers, ice skaters, rugby players, ice hockey goalkeepers, tennis players, basketball players and the rest, the body will create appropriate skill sets if exposed enough to the specific demands of the sport. By training on a regular basis, a person's functional capacity will develop to match the demands placed upon the body. So maybe my daughter would have coped well with being placed in a point of pain when stretching at her local gymnastics club and would have adapted well to these demands, but would it have been worth the risk? The good news is that national governing bodies for various sports are recognizing that there need to be guidelines for coaches, parents and participants to ensure that the participants experience a long and healthy time involved in sport and activity, and that the risk of injury is minimized.

POST-OPERATIVE AND CHRONIC PAIN: STRETCHING, MOBILITY AND REGAINING FUNCTION

So what about post-operative stretching? Or chronic pain? Can an increased ROM and improved mobility help at all?

The body's response to surgery is the same as to an acute injury, treating it as an injured/damaged site and leading to increased swelling, bleeding into the area and reduced ROM, whether something has been removed, replaced or repaired in the surgical area. The healing process will be the same as with an acute injury: the destruction/inflammatory response phase; the proliferative/regeneration phase; and the remodelling phase. Having reviewed a number of post-surgical guidelines from surgeons online, all of them mention the importance of regaining full functional ROM. Again, the process is static-active stretching progressing to static-passive stretching relative to the operative site and associated tissues. This will also help with improved circulation, pain management and strength gains over time.

Ayhan et al.[108] analysed the effect of neck-stretching exercises, following a total thyroidectomy, on reducing neck pain and disability. The participants were randomly assigned either to the stretching exercise group (n = 40) or to the control group (n = 40). The stretching exercise group learnt the neck-stretching exercises immediately after the total thyroidectomy. The effects of the stretching exercises on the participants' neck pain and disability, neck sensitivity, pain with neck movements, as well as on wound healing, were evaluated at the end of the first week

and at one month following surgery. Ayhan et al. found that when comparing neck pain and disability scale scores, neck sensitivity and pain with neck movement before thyroidectomy, after one week patients experienced significantly less pain and disability in the stretching exercise group than in the control group. However, they found that there was no significant difference between the groups' scores at the one-month evaluation. Ayhan et al. conclude that neck-stretching exercises done immediately after a total thyroidectomy reduce short-term neck pain and disability symptoms, but that there is no significant change long term between the stretch group and the non-stretch group.

However, regaining full ROM is essential for all post-operative interventions, from a mastectomy, knee replacement, thyroidectomy, shoulder surgery, abdominal surgery – just anywhere! Educating the patient to understand the importance of early mobilization is essential.

The positive outcome for the client in the case study is fantastic – it is why I do what I do! It demonstrates what a dedicated mobility and strengthening programme can achieve for chronic pain. It is a long-term commitment, though, and as we saw with yoga and Pilates, gains can quickly disappear with detraining and non-adherence to the plan.

Lee et al.[109] investigated whether a stretching programme reduced acute musculo-skeletal impairments in patients undergoing radiotherapy for breast cancer post-surgery. Often a loss of strength and ROM in the upper quadrant is experienced in breast cancer patients, so could a stretching programme help to reduce this loss? The control group received no advice about exercise. The stretch group received instruction on low-load, prolonged pectoral stretches, which were to be performed daily and were checked at weekly visits. Shoulder ROM, strength, arm

CASE STUDY: MASTECTOMY WITH DIEP RECONSTRUCTION

This case study is of a female client, aged fifty-four, who in autumn 2009 underwent a mastectomy, followed by DIEP flap reconstruction (in which fat, skin and blood vessels are cut from the wall of the lower belly and moved up to the chest to rebuild the breast). She presented to me five years later with post-surgical abdominal pain that had been on-going and constant.

Goal: to reduce pain in stomach/abdominal area.

Prescribed plan: to improve core and spinal strength, improve movement patterning and introduce regular stretching and improved mobility.

Outcome: good long-term adaptation and positive change in pain status.

Client Testimonial

A friend recommended Alex to me as I suffer from abdominal pains after a mastectomy and DIEP reconstruction using tissue from my abdomen. The operation was five years ago and I have been left with at best a constant feeling of heaviness/tightness. However, in January 2014 this worsened and I was in constant pain with a darting, tingling pain. The surgeon has explained that this can sometimes happen and there is nothing that they can do medically (well, I suppose apart from endless pain killers) to deal with it.

I have to admit that I wasn't sure about going to Alex – I'm not at all athletic and hate exercises – and I'm quite pedantic about understanding what I am doing and ensuring that the movements I'm making are correct. Also I have to master one aspect before I can move on to the next – so showing me an exercise and then asking me to breathe at the same time proved tricky at times. I have been through a few Pilates teachers who lacked the stamina to cope with me.

I'm delighted to say that Alex not only has stamina but also patience and diplomacy skills to see even someone like me through an exercise programme. When I whinged about the exercises being painful we stopped and discussed my motivation and how important was it for me to reduce the pain I was in. She convinced me that even, if at worst the pain did not subside, the core area we were addressing would be stronger to cope and that overall my posture and fitness would improve – a positive thing in itself.

Even after the first week I felt a benefit and of course success breeds success and the better I felt the more inclined I was to believe in and do the exercises. Whilst not out of pain, it has certainly decreased and not only do I feel physically better but I'm also mentally better at dealing with it.

I keep a daily score sheet of what exercises I have done (and how many) which I show to Alex and it helps to remind me to query if there is some aspect of the exercise I don't feel I'm doing right or if I felt particular pain in one. Alex is always very encouraging when she looks at it – and her positive comments keep one motivated.

I think one of the main differences with Alex (compared to the Pilates/exercise people I had before) is that because she does rehabilitation she is used to dealing with people in pain and understanding their needs rather than a regimented fitness programme with a 'one size fits all' approach. She cares about the individual and wants to make us better.

Anyway I'm absolutely delighted to find that a fitness programme can help with my pain – much to my amazement – and that it just gives me such a lift. Thank you Alex.

circumference and quality of life measurements were taken prior to, and at completion of, radiotherapy and at seven months after radiotherapy.

The conclusion of the review was that there was no difference in any outcome between the two groups. Breast symptoms increased for both groups during radiotherapy, without loss of strength or ROM. The incidence of lymphedema during the study was low for both groups and did not differ between them. The pectoral stretching programme did not influence the outcomes measured, because the symptoms reported by patients were not a consequence of contracture, so there was no loss of ROM as a result of the surgery or treatment in this group. I wonder whether there was a benefit in the patient's self-esteem from completing the stretches during treatment? However, quality of life measurements were taken and there was no difference with either group.

Chronic low back pain is unfortunately a common occurrence and around 80 per cent of people will experience this annoying and painful condition at some time in their lives. Can stretching and resultant improved ROM and mobility help to prevent or alleviate this pain? Or just strength and core work? Or a combination of both? Castagnoli et al.[110] compared global postural re-education (GPR) to a standard physiotherapy treatment based on active exercises, stretching and massaging for improving pain and function in chronic low back pain patients. Their clinical study aimed to describe the short- and long-term effects of GPR on patients with non-specific chronic low back pain, compared with a similar sample of control that received standard individual physiotherapy treatments. The study group consisted of adult patients with a diagnosis of non-specific, chronic (more than six months) low back pain. Interventions consisted of fifteen sessions of one hour each,

twice a week, including patient education for both groups.

GPR is an approach based on an integrated idea that the muscular system is formed by muscle chains, which can face shortening as a result of constitutional, behavioural and psychological factors. The aim of GPR is to stretch the shortened muscles using the creep property of viscoelastic tissue and to enhance the contraction of the antagonist muscles. The participants all completed two questionnaires at baseline, discharge, fifteen working days from baseline and at twelve months from discharge. The primary outcome measure was low back pain-related functional disability, assessed by the Roland and Morris Disability Questionnaire in addition to the Numeric Rating Scale (NRS), which consists of a line of numbers from 0 to 10, representing pain severity levels from 'none' to 'most intense pain imaginable'. No clinical assessments were completed to assess improvement in the participants, only the questionnaires.

The results in both groups and for both outcomes showed that there was a statistically significant improvement in Roland and Morris Disability Questionnaire and NRS scores on discharge. One year after discharge, they found an improvement in both the Roland and Morris Disability Questionnaire and NRS compared to baseline, but only pain relief, as expressed by NRS improvement, was statistically, but not clinically, significant in the GPR group. GPR patients reported similar improvement in pain and function as those who received standard physical therapy in the short term, as both treatments were associated with statistically significant improvements in function and pain, while only GPR treatment was associated with statistically significant pain relief at the one-year follow-up. Physiotherapists need to attend additional training to be able to practise GPR, so there is a cost implication regarding this. Stretching is a normal treatment method for patients with chronic low back pain, in addition to strengthening and education, so the addition of GPR, according to Castagnoli et al., may not be warranted as it elicits similar outcomes as traditional physiotherapy.

Cunha et al.[111] also used the application of GPR this time in chronic neck pain. They compared the effect of conventional static stretching and muscle chain stretching, as proposed by the GPR method, in the manual therapy of patients with chronic neck pain. Similarly to Castagnoli's studies, it was concluded that conventional stretching and muscle chain stretching in association with manual therapy were equally effective in reducing pain and improving the ROM and quality of life of female patients with chronic neck pain, both immediately after treatment and at a six-week follow-up, suggesting that stretching exercises should be prescribed to chronic neck pain patients. The types and application of stretching for the two groups were, first, GPR posterior muscle chain stretching and GPR anterior muscle chain stretching. The two postures were held for 15min each. Gradually, respecting the patient's limits, the lower limbs were extended as much as possible throughout the duration.

The second type was conventional static stretching, consisting of upper trapezius, suboccipitalis and back of the neck, pectoralis major and minor, rhomboids, finger and wrist flexors, forearm pronators, finger and wrist extensors, forearm supinators and paravertebral muscles. Each exercise was repeated twice for 30sec and done slowly at normal breathing rhythm. Both stretching modalities resulted in the same duration of stretching time.

Global Postural Re-education posterior chain.

Global Postural Re-education anterior chain.

DEFINITION OF CONTRACTURES

Contractures are the chronic loss of joint motion due to structural changes in non-bony tissue. These non-bony tissues include muscles, ligaments and tendons. Contractures can occur at any joint of the body. This joint dysfunction may be a result of immobilization from injury or disease; nerve injury, such as spinal cord damage and stroke; or muscle, tendon, or ligament disease. There are a number of pathologies and diseases that can lead to joint contractures. The primary causes resulting in a joint contraction are muscle imbalance, pain, prolonged bed rest and immobilization. Because of the frequency of fractures and surgery, immobilization is the most frequent cause of joint contractures. Symptoms include a significant loss of motion to any specific joint that results in immobility. If the contracture is of a significant degree, pain can result even without any voluntary joint movement.

(www.thefreedictionary.com)

Cunha *et al.* conclude that conventional stretching and muscle chain stretching, in association with manual therapy, were equally effective in reducing pain and improving ROM and quality of life in female patients with chronic neck pain, both immediately after treatment and at a follow-up six weeks later. Since muscle stretching is a low-cost treatment, it should be pursued more often for treating chronic neck pain.

So it appears that stretching and, specifically, static stretching in active and passive modes can help with post-operative and chronic pain issues; even if this is anecdotal, it is valuable for the patient. Increased ROM in off-loading irritable joints, releasing scar tissue or adhesions, improved stretch tolerance and a return to normal, functional ROM may all go towards the reduction in pain for individuals.

The benefit of static stretching is that it is easy to control and apply in any setting, so as long as the individual adheres to the prescribed plan, there should be a positive outcome. Long-term gains are not mentioned in the above studies, as they involved only six-month and one-year reviews, but I would assume that in a chronic problem the individual would need to commit to long-term adherence to the plan in order to maintain and improve flexibility and mobility, as well as any appropriate strength work, so as to maintain a positive long-term outcome.

Another chronic problem that affects many people is contractures. Can a dedicated stretching programme help the affected area? Katalinic *et al.*[112] wanted to determine the effects of stretch on contractures in people with, or at risk of, contractures. Contractures are a common complication of neurological and musculoskeletal conditions and are characterized by a reduction in joint mobility. Stretch is widely used for their treatment and prevention. The review by Katalinic *et al.* included thirty-five different studies and they concluded that in people with neurological conditions, there was moderate to high-quality evidence to indicate that stretch does not have clinically important immediate, short-term or long-term effects on joint mobility for this cohort. This is certainly an area that may require further investigation.

Post-Operative and Chronic Pain Application

Mode: static-active stretching progressing to static-passive stretching relative to post-operative site and/or associated area/tissues in chronic pain. GPR has been shown to have a similar outcome to traditional conventional stretching, but the practitioner needs to have additional training to apply this form of stretching.

Rationale: to improve post-surgical ROM. ROM will be reduced as a result of swelling, bleeding and scar tissue formation around the post-operative site, so a normal functional ROM needs to be regained. The aim is management of chronic pain, with improved ROM and stretch tolerance and a return to normal function, when used in combination with appropriate manual therapy.

Outcome: improved ROM and return to pre-operative functional ROM; reduction in chronic pain.

Sets: as clinically appropriate: two to three sets within a pain-free ROM.

Duration: 30sec static holds.

Frequency: as clinically appropriate; certainly regular/daily exposure.

Long-term benefit: to return to pre-surgical ROM and full function; reduction in chronic pain and improved self-esteem.

ADVERSE NEURAL TENSION

One of the areas of clinical practice that chartered physiotherapists, athletic trainers and sports medicine practitioners commonly assess is neural tension. A musculoskeletal assessment will look at ROM through the skeletal system, but it is also important to consider the effect that the nervous system has on people. Does a positive slump test affect ROM and injury potential?

McHugh et al.[113] looked at the role of neural tension in hamstring flexibility to determine if neural tension, via the slump test during passive hamstring stretch, affected maximum ROM, stretch discomfort, resistance to stretch, or the electromyography (EMG) response to stretch in healthy subjects. They hypothesized that neural tension would decrease maximum ROM and increase stretch discomfort without affecting the EMG response or resistance to stretch.

Resistance to stretch, knee flexion ROM, hamstring EMG activity and stretch discomfort were measured during passive hamstring stretches in eight subjects. The subjects had no significant history of low back or hamstring injury, were in good general health and were active.

Stretches were performed with the thoracic and cervical spine in neutral position (neutral stretch) and in the slump test position (neural tension stretch). Stretches were performed using an isokinetic dynamometer (5 degrees/sec) with subjects seated, the test thigh flexed 40 degrees above the horizontal and the seat back at 90 degrees to the horizontal. Knees were passively extended from 90 degrees flexion to maximum stretch tolerance. Subjects held a switch in their hand during stretches that automatically stopped the stretch at a self-determined maximum ROM. The stretch was then released. For neural tension stretches, the cervical and upper thoracic spine were manually flexed by a physical therapist to illicit a neural response. Subjects were asked to grade their stretch discomfort on a scale of 0 to 10, where 0 equalled 'no stretch discomfort at all' and 10 equalled 'the maximum imaginable stretch discomfort'. The primary finding in this study was that adding neural tension during a hamstring stretch (slump position) increased the resistance to stretch compared with a hamstring stretch with the spine in a neutral position.

McHugh et al. discuss and conclude that this increased tension may be due to resistance to elongation of the neural tissues, or adhesions between the neural tissues and the surrounding tissues that prevent the neural tissues from gliding freely, as previously discussed in our discussion of slump assessments earlier. The clinical and practical relevance of these findings with McHugh's study are twofold:

■ Adverse neural tension may be a contributing factor in the high recurrence of hamstring strains (as an indicator of adhesions within the muscle or as a source of recurrent pain).
■ Neural tension during passive stretching might be the neural mechanism for stretch-induced strength loss. The role of neural tension in hamstring strain recurrence and stretch-induced strength loss can be examined in future studies using this methodology.

So should the prescription of slumps to help relieve adverse neural tension be in the conditioning plan for those individuals who have experienced a hamstring strain?

Indeed, Kornberg and Lew[114] looked at a group of professional Australian Rules football players. This study was conducted on players diagnosed as having grade I hamstring injuries, who demonstrated a positive response to the slump test. Of the twenty-eight subjects that satisfied the inclusion criteria, sixteen were treated traditionally, with the remaining twelve receiving slump stretch as an addition to the traditional treatment regime. The results indicated that traditional treatment plus slump stretch technique was more effective in returning the player to full function than the traditional regime alone. Therefore, the authors suggest that the slump test should be a mandatory test in the assessment of hamstring strain, so that the more effective treatment regime may be implemented.

Turl and George[115] also reviewed the effect of neural tension in relation to athletes with a previous grade I hamstring strain. They investigated the presence of adverse neural tension in fourteen male Rugby Union players with a history of grade I repetitive hamstring strain. A comparison was made to an injury-free matched control group. Adverse neural tension was assessed using the slump test. Hamstring flexibility was measured using the active knee extension in lying test. Results indicated that 57 per cent of the test (previous history) group had positive slump tests, suggesting the presence of adverse neural tension. None of the control group had a positive slump test. Analysis of variance revealed no differences in flexibility between groups, or between those demonstrating a positive or negative slump test. These results also suggest that adverse neural tension may result from, or be a contributing factor in, the aetiology of repetitive hamstring strain.

Certainly, more research is needed in this area, but it makes a lot of sense, regarding the challenge for athletes and practitioners in the management of repeated hamstring injuries, to include neural mobility and gliding within a rehabilitation and conditioning programme. I prescribe it for many of my athletes, who tend to be in a flexed position when driving and sitting at a desk and who are physically active, with most having a positive slump test. I prescribe neural slumps as part of their warm-up and activation routine to mobilize any adhesions. They are educated in the concept and also warned that they could irritate the nervous system if they complete too many or too vigorously, but they appear to experience positive physical gains by its inclusion. This is perhaps an 'experience' decision by myself and our chartered physiotherapist, who also supports its inclusion, rather than a research-based decision, but from the evidence above we know that ROM will be compromised with adverse neural tension and that a positive slump may be present in those with previous hamstring injury, so this may therefore be good preventative practice.

Adverse Neural Tension Application

Mode: slump position as completed in a neural tension assessment.

Rationale: to reduce neural adhesions and encourage gliding through the neural system. To be completed by those with a positive slump test.

Outcome: positive slump test progresses to a negative slump test and full pain-free ROM with zero neural tension.

Sets, repetitions and frequency: as clinically appropriate. Three sets of eight to twelve repetitions left and right side in a slump position; alternate sides. Completed before exercise and also possibly on rest days.

Long-term benefit: to mobilize adhesions within the neural tissue to help reduce neural symptoms and improve reduced ROM. Important post-hamstring strain to reduce incidence of re-injury.

PSYCHOLOGICAL FACTORS AND ROUTINE IN RELATION TO STRETCHING

It is important to recognize that despite all the research-based evidence referenced within this book, guiding people into best practice to ensure an optimal outcome for their individual circumstances in relation to stretching may still be a challenge.

Some people will still want to perform the same routine and same stretches that they have always done, despite any possible contraindication and despite the latest research, which may imply a reduction in performance measures. As Paul mentioned in the Foreword, if a player has had a bad game after they have been introduced to a new stretching concept or training mode, then typically the player or athlete will not be doing that again, no matter how good it may be for them. On the flip side, if they had an amazing performance, they will almost certainly buy into the new regime and plan!

The psychological component of what choices we make is important to us and of course should be respected. Our job as professional practitioners is to educate and guide our clients and athletes to ensure optimal performance, but sometimes your athlete will just do what they want to do! Chris Bodman, who is a sports psychologist for a number of professional teams and elite sporting clients, discusses the rationale of this in relation to psychology and performance choices.

As Chris mentions in the Feature Box, stretching and habitual routines as an individual or within a team or group are important to sports people and can become ritualistic. Even if someone is told that static stretching, for example, may not have any worth prior to an event or activity, the individual may still want to complete some static stretches as part of their routine. It is important to respect and recognize this. The same applies to DOMS and post-session recovery techniques, where an individual may feel the benefit psychologically from static stretching post-session, even though the research may contradict this behaviour. Education as to best practice is always preferential in leading to the best performance outcome and should be encouraged. However, sometimes it is also okay to respect how a person still chooses to do something.

PSYCHOLOGY OF PERFORMANCE CHOICES BY CHRIS BODMAN MBPSS

The pursuit of flexibility has been a key aim of performance training since people began exercising. Yet how flexibility relates to performance and injury prevention often remains inconclusive or ambiguous, as we have found out within this resource. Although some research provides opposition to the physiological effects of stretching, there remain theories which contend that stretching may provide psychological benefits.

As with any intervention, a specific needs analysis of the sport, position and individual athlete is imperative. For some, stretching is not necessarily the most comfortable pastime in the world. If individuals find it intolerably painful, then they may not do it regularly. Each person is different and it is important to understand the point at which they are comfortable to stretch. An important psychological point here is that the body does not like to be moved too far out of its comfort zone. Therefore, muscles that have been short and tight for a long time are going to need to be convinced to change their length by both consistent application and possibly other methods that will override the body's drive to remain in its comfort zone.

Stretching is often part of an athlete's routine. For many athletes, or players of team sports, stretching before an event may allow them to focus on the upcoming task demands, making them feel more prepared to perform.

Stretching can also be a way of reducing anxiety and finding optimal states of arousal. In this case, the benefits of having the right psychological mindset may negate some of the inconclusive performance and injury-prevention measures mentioned here. In fact, stretching regularly may even promote an increased awareness of other factors. For example, allowing an athlete to become more attuned to muscle soreness may encourage them to take appropriate measures to alleviate that soreness before competition. In team sports, a collective stretching routine is commonplace. Again, in many instances this may not have a significant effect on injury prevention or performance enhancement, but may appear a more habitual, ritualistic activity. However, it often provides both a sense of team unity and a belief and commitment that stretching has a positive impact. This may be enough for athletes to become more mindful of maintaining good health, allowing them to compete at a higher level for longer.

Stretching after an event may be beneficial to athlete psychological recovery and perception of pain. Whilst some of these subjective measures may not be in line with other physiological markers, they are nonetheless important in interpreting an athlete's psychological recovery. Even if stretching does not influence objective measures of muscle damage, athletes who feel better may gain psychological benefits and be more motivated for future training and competition. Although the literature suggests that the acute effects of stretching may show a decrease in performance, if the majority of flexibility training is at the end of the day, after competition or practice is over, athletes can gain the long-term benefits of stretching without compromising immediate performance on the field.

Stretching after an event may also allow athletes time to reflect on their performance, either individually or collectively. It may also be used as an alternative relaxation technique in order to lower states of arousal. As previously mentioned, the individual approach is of utmost importance. Whilst some athletes may choose to stretch immediately after an event, others may choose to stretch

(continued)

(continued)

at home as a daily routine, or through yoga classes or other structured formats. Allowing for these preferences will again be more psychologically advantageous for an athlete.

Although more research is welcomed to fully understand the psychological benefits of stretching, there appears to be enough anecdotal evidence to suggest that there are both short-term and long-term benefits of a personalized approach.

Group stretching session: psychological gain, physiological gain, or social interaction?

EXERCISE-ASSOCIATED MUSCLE CRAMP AND SLEEP-RELATED MOVEMENT DISORDERS

Muscle cramp during physical exercise and pregnancy, as well as sleep-related movement disorders such as nocturnal cramps and restless leg syndrome, can be painful, annoying and frequent for some people.

A cramp is a painful, involuntary and spasmodic contraction of a muscle. It is important that the cause of the cramp is not due to an underlying serious clinical condition, so if cramps occur on a regular basis, medical advice certainly should be sought. In the instance of cramp, stretching is often prescribed to help relieve the sensation, but does this help?

Miller et al.[116] analysed the research available between 1955 and 2008 in regard to exercise-associated muscle cramps. Cramp is a common condition experienced by recreational and competitive athletes, and can be identified when the muscle goes into a spasm and causes acute local muscular pain, stiffness, visible bulging or knotting of the muscle and possible soreness that can last for several days. They discuss two main theories that may cause cramps in this population: the dehydration-electrolyte imbalance theory; and the neuromuscular theory. Although the cause is not certain, it appears that the research points more towards the neuromuscular theory, which is that the muscle is overloaded beyond its capacity and neuromuscular fatigue causes an imbalance between excitatory impulses from muscle spindles and inhibitory impulses from Golgi tendon organs. If the dehydration-electrolyte theory was valid, they discuss, then by purely increasing fluid intake and appropriate electrolyte intake this would prevent cramping, which does not happen. According to Miller et al., there may be numerous other variables connected to exercise-associated cramps that could be factors, for example, accumulation of metabolites, intensity of exercise, level of conditioning and heat acclimatization.

So prevention may be possible by remaining well hydrated and well-conditioned and acclimatized to the environment, but if an exercise-associated muscle cramp occurs, what is the best mode of treatment? A number of papers have discussed the benefit of static stretching as a result of the onset of exercise-associated muscle cramps.

Miller et al.[117] once again looked into this subject. This time, they considered that static stretching pre-exercise is prescribed to prevent cramps based on the assumption that Golgi tendon organ inhibition remains elevated post-stretching. Their research determined whether stretching increased gastrocnemius Golgi tendon organ inhibition and, if so, what was the time course of this inhibition post-stretching. We know from our review on static stretching that the tissue returns back to pre-stretch level within 3min of 30sec holds (DePino et al.),[118] but what Miller et al. considered was an additional transcutaneous stimulation and longer static holds. The investigators applied 3 × 1min duration gastrocnemius stretches.

Participants maintained voluntary contraction intensities of 5 per cent of their maximum, while the Achilles tendon was stimulated transcutaneously fifty times. Five-hundred millisecond epochs of raw electromyographic activity were band-pass filtered, full-wave rectified and averaged through the Achilles. This research by Miller et al. indicates that pre-exercise stretching does not affect the Golgi tendon organs beyond the 30min stretching period, so if pre-stretching does prevent fatigue-induced cramping, the mechanism is unlikely to involve the auto-inhibition produced by the Golgi tendon organ reflex.

Most people do not stretch prior to exercise to prevent cramping, unless they perhaps have a history of regular cramping. Pre-exercise stretching is generally conducted to prepare for activity and it is more likely that someone will stretch in a reactive manner as the involuntary and spasmodic contraction occurs.

Bentley[119] discusses the neural component of cramps and how extreme fatigue and other factors may cause exercise-associated muscle cramps. Disturbances at various levels of the central and peripheral nervous system, and skeletal muscle are likely to be involved in the mechanism of cramp and may explain the diverse range of conditions in which cramp occurs. There are no proven strategies for the prevention of exercise-induced muscle cramp, but regular muscle stretching using post-isometric relaxation techniques, correction of muscle balance and posture, adequate

Reactive stretching as a result of cramp.

conditioning for the activity, mental preparation for competition and avoiding provocative drugs may be beneficial. Other strategies such as incorporating plyometrics or eccentric muscle strengthening into training programmes, maintaining adequate carbohydrate reserves during competition, or treating myofascial trigger points are speculative and require investigation.

Certainly the use of medications like quinine, for example, seem to be discouraged by medical experts. Stretching is a non-invasive, non-harmful reactive solution – we all seem instinctively to start to stretch as we reach for the cramping muscle and roll around in pain! However, perhaps more progressive for chronic cramp sufferers is the research by Behringer et al.,[120] who investigated if the cramp threshold frequency (CTF) could be altered by electrical muscle stimula-

tion in a shortened muscular position. A total of fifteen healthy male sport students were randomly allocated to an intervention group (n = 10) and a non-treatment control group (n = 5). Calf muscles of both legs in the intervention group were stimulated equally twice a week over six weeks. The protocol was 3 × 5sec on, 10sec off, 150µs impulse width, 30Hz above the individual CTF, and was at 85 per cent of the maximal tolerated stimulation energy. One leg was stimulated in a shortened position, inducing muscle cramps, while the opposite leg was fixated in a neutral position at the ankle, hindering muscle cramps. CTF tests were performed prior to the first and ninety-sixth hour after the sixth (three weeks) and twelfth (six weeks) training session. They discovered that, after three weeks, the CTF had significantly increased in the treated calves of the intervention group, while it remained unchanged in the non-treatment control group and non-stimulated calf in the intervention group. They conclude that the present study may be useful for developing new non-pharmacological strategies to reduce cramp susceptibility.

The concept of training and conditioning the muscles to increase this cramp threshold seems to lead to a successful adaptation in a reduction of cramps, which could be a positive non-invasive, medication-free intervention for those who struggle with cramp. Research into nocturnal cramping with this modality may be warranted and valuable.

Hallegraeff et al.[121] completed a randomized trial of whether stretching before sleep reduces the frequency and severity of nocturnal leg cramps in older adults. Their sample included eighty adults aged over fifty-five years with nocturnal leg cramps who were not being treated with quinine, which is often prescribed to help with nocturnal cramps.

The experimental group performed stretches of the calf and hamstring muscles nightly, immediately before going to sleep, for six weeks. The control group performed no specific stretching exercises. Both groups continued other usual activities. The stretching group performed standing calf and hamstring stretches for 3 × 10sec duration, with 10sec of relaxation between each stretch. Stretching of both legs was done within 3min. The participants were advised to adopt the stretch position and move to a comfortable ROM, move beyond this until a moderately intense stretch was felt and sustained for 10sec, then return to the starting position.

The results at six weeks indicated that the frequency of nocturnal leg cramps had decreased significantly more in the experimental group, with a mean difference 1.2 cramps per night. The severity of the nocturnal leg cramps had also decreased significantly more in the experimental group than in the control group, with a mean difference 1.3 on the 10 visual analogue scale. The best estimate of the average effect of stretching on the frequency of nocturnal cramps was a reduction of about one cramp per night. Given that the participants had an average of approximately three cramps per night at the beginning of the study, this is a beneficial and easily applied mode of treatment.

Contrary to Hallegraeff's paper, Coppin et al.[122] assessed the effect of calf-stretching exercises and cessation of quinine treatment for patients with night cramps. They concluded that calf-stretching exercises are not effective in reducing the frequency or severity of night cramps. At twelve weeks, there was no significant difference in the number of cramps in the previous four weeks, nor a reduction in symptom burden or severity of cramps.

Garrison[123] questioned the sample population from the research completed by Hallegraeff et al., as their sample group appeared to have significantly higher rates of nocturnal cramp than other studies. Garrison therefore questioned the objectivity of the study. He believes that the current body of evidence does not support bedtime stretching for the prophylaxis of nocturnal leg cramps.

This is certainly a confusing and vague area and undoubtedly more research is needed to answer these questions. However, if an individual feels better and may have a subsequent reduction in nocturnal cramping as a result of a short bout of calf and hamstring stretching before bedtime, then surely it is a potentially helpful technique and certainly does not have any harmful side effects.

Sleep-related movement disorders, such as restless leg syndrome, can occur during long-haul flights, or when sitting still for a long time, for example. Symptoms include an urge to move the legs, usually accompanied by an uncomfortable sensation in them. Suzuki et *al.*[124] report an improvement of symptoms after movement, such as walking and stretching. They recognize that symptoms occur or worsen during periods of rest and in the evening and night. Stretching has historically been prescribed, along with dynamic movement to help increase circulation and relieve symptoms, for example on a long-haul flight, travel or sitting for extended periods of time. The causes of restless leg syndrome are vague, but, again, prophylactic stretching appears to be beneficial.

CHAPTER 6

PUTTING IT ALL TOGETHER: APPLICATION IN THE PRACTICAL SETTING

I hope that the applied principles and literature summarized above will provide a valuable resource, which will allow you to consider your stretching methodology and its health- and performance-related benefits and subsequent appropriate application. The research is still in its infancy and there are often inconsistencies, or reasons why a study may or may not be absolutely conclusive regarding one mode of stretching or another. My aim has been to provide a research-based summary of many of the different modes and types of stretching, so as to allow an informed decision to be made. I always reflect on why I have done or prescribed something and what the outcome has been – this is how we learn and gain experience and ensure best practice. I am hoping that the insight this book provides will lead to better, more appropriate prescription of the different stretching modalities by practitioners, coaches, athletes and the general population.

All modalities have benefits, but some benefit a sedentary population more than an elite athlete and vice versa. Knowing this and becoming more aware of the appropriate application is important for the relevant health and performance gains of the individual. Franco et al.[125] examined the acute effects of different stretching exercises on the per-

formance of the traditional Wingate test. The Wingate test is a common dynamic test used to evaluate an athlete's anaerobic performance. Fifteen male participants performed five Wingate tests: one for familiarization; and the remaining four after no stretching; static stretching (3 × 30sec static holds); dynamic stretching (exercise consisting of three sets of five slow repetitions followed by ten fast repetitions completed as fast as possible); and PNF. The PNF exercise was performed three times, with the participant achieving maximum tolerable ROM of the targeted muscle while an experimenter provided an opposing force for 8sec, followed by relaxation. Stretches were targeted for the hamstrings, quadriceps and calf muscles. Peak power, mean power and the time to reach peak power were calculated.

The results were interesting, demonstrating that the time to peak power presented the most consistent pattern in terms of differences across stretching conditions, because a consistent delay of this peak was observed after all stretching exercises. This meant that there was a delay in the ability to reach peak power after all stretching protocols, compared to the familiarization test, and that non-stretching achieved the fastest time to peak power. Non-stretching resulted in the highest mean power value, with mean power values higher in dynamic stretching, then reduced for static stretching and finally lowest in PNF. Peak power was higher when the group performed dynamic stretching, then the non-stretching group, static stretching and finally PNF.

The main findings of the investigation by Franco et al. were that stretching decreased performance by lowering peak power and that time to achieve peak power on a Wingate test was delayed. A consistent increase of time to reach peak power was observed after all stretching exercises when compared to non-stretching. Franco et al. conclude that the type of stretching, or no stretching, should be considered by those who seek higher performance, or practise sports that use maximal anaerobic power. This paper provides an example of a 'summary' of stretching in relation to power and demonstrates the kind of considerations and challenges that we have when deciding whether to stretch or not to stretch.

In performance, this has been a very common theme within the research. Performance measures only appear to be enhanced with dynamic and ballistic stretching before exercise, but it is unclear as to whether this is as a result of the specific stretching modality, or of the sport-specific warm-up completed on its own or in combination with the stretching modality.

Certainly, ROM can be improved in the long term with many of the stretching modalities discussed here. Some modalities have a minimal or no effect on performance measures, like myofascial release, ballistic stretching, dynamic stretching or vibration training, so can be performed before exercise to increase ROM. Yet others, like static stretching and PNF, have been shown to reduce performance measures and physical capacity in the form of strength and power output, so

should be avoided before exercise or competition.

Acute injury or post-operative static stretching has been demonstrated to help regarding regaining a normal ROM in the damaged limb and it is important to recognize how early mobilization can help with a return to full function after surgery or an injury, bearing in mind the healing time and properties of the muscles and any other relevant clinical detail.

Increasing joint ROM in young people who participate in sports has been shown to reduce the incidence of growth injuries, such as Osgood-Schlatter disease and Sever's disease, and also in chronic injuries, such as plantar fasciitis. So an effective stretching programme, focusing on increased ROM in this cohort, may be beneficial relating to injury-prevention strategies and should be included in training.

Functional ROM occurs as a normal training adaptation in sports people, so that they become conditioned to complete movement patterns within the demands of their sport or activity. It is when someone is unconditioned or unaccustomed to a movement pattern or ROM that injury may occur. Neural stretching in the form of slumps has been raised as an important ingredient within a return to training plan for those who have already experienced hamstring strains and may be beneficial as part of a warm-up/activation programme to release neural adhesions to prevent injury before exercise.

Exercise class programmes, such as yoga, Pilates and static-stretching groups, have demonstrated positive outcomes, especially in sedentary or elderly populations, where significant physical improvements have been observed. Also, prophylactic stretching appears to be beneficial for people who experience cramp or sleep-related movement disorders and is a reactive solution to exercise-associated muscle cramps. More research is certainly required in this area, which affects many people.

There are a number of screens or assessments that can evaluate ROM for an individual and assess if movement is compromised. These are important when considering the effectiveness of a prescribed stretching programme to observe any changes and also after acute injury, surgery or chronic injury to regain normal ROM. The tests need to be valid, reliable, objective and repeatable.

There is much information here and, as with all research-based resources, there will be limitations and variations within the studies

cited, where the ability to compare so many different studies and come to any valuable conclusion is of course a challenge. Hopefully, the summary at the end of each modality will be useful. New research will arise over time and it is certainly needed, as it appears that there are positives and negatives and different outcomes with each possible option. Making an informed decision regarding best practice is important and, as discussed, will affect the performance outcomes.

The summary table below provides a quick visual regarding which stretching technique will provide a specific outcome and what the applied prescription should be.

Summary				
Stretching Modality or Presentation	Rational and Outcome	Prescription	Population	DOMS Benefit?
Static	May have positive gains on a number of performance variables for sedentary populations. No performance gains with acute application before exercise and even a reduction in performance measures (strength and power output).	>30sec holds 3 × week. 2–3 sets per body part.	Sedentary or older populations.	Nil
Dynamic	Warm-up prior to a sporting event or before physical activity. Enhances muscular performance and flexibility/ROM before exercise.	20–30sec per body part. 1–2 sets per body part. 6–8min prior to exercise × 2 blocks if desired. Research indicates that 1 set appears to be enough for performance gains. Complete prior to all training and sporting events.	All active populations.	Nil

Stretching Modality or Presentation	Rational and Outcome	Prescription	Population	DOMS Benefit?
Ballistic	Sport-specific explosive warm-up prior to a sporting event, or before physical activity that demands ballistic/ explosive movements. Increased ROM/ flexibility. Improved vertical jump.	30sec duration on each side for each limb. 2 sets on all selected muscle groups. Metronome beat around 60 beats/minute. Before sports events and before training so to be conditioned/prepared for sport-specific movement patterns.	Active and conditioned populations.	Nil
PNF & Isometric Holds	Rehabilitation setting or when increased ROM is desired. Improved ROM, with no subsequent effect on strength or power. May affect anaerobic peak power (Wingate test).	3–4 reps per limb per set. 1–2 sets per limb. 6–10sec isometric holds per repetition. Contraction intensity: between 20–100%. Optimal: between 60–100% 3 × each week.	All populations, especially appropriate for rehabilitation.	Nil
Myofascial Release	Recovery from DOMS. Pre- and post-exercise application for increased ROM, with no subsequent performance losses. Improved recovery with reduction in pain from DOMS, increased ROM and also improved performance indicators post-high intensity exercise.	1–2 reps per muscle group. Movement cadence: every 1sec or so on an area working up and down particular muscle group. 45sec moving up and down the muscle, per muscle group. 15sec rest between each muscle group. Pre-exercise for improved ROM and immediately post-exercise and 24hr, 48hr after intense sessions to help reduce DOMS.	All populations.	Positive outcome with myofascial release.

Stretching Modality or Presentation	Rational and Outcome	Prescription	Population	DOMS Benefit?
Whole-Body Vibration Training	Increased ROM/ flexibility in addition to improved performance measures. Flexibility improvements/ increased ROM. Increased muscle activation to improve performance measures. And also studies imply improved health-related benefits, like improved body composition and bone density in post-menopausal women, as well as strength gains/ improvements.	1–2 reps per muscle group/limb. Hertz: range from 30–50Hz. Amplitude: 2–4mm. Duration: 20–60sec per limb/muscle group. 15sec rest between each muscle group 3 × each week.	All populations; although more research needed.	Nil
Pilates	Body conditioning and muscular endurance, rehabilitation, postural and spinal issues/ rehabilitation and flexibility/mobility. Improved flexibility, abdominal endurance, hamstring flexibility, upper-body muscular endurance measures after regular attendance.	2–3 sessions per week. 45–60min classes. Long-term benefit: research demonstrating benefits from 8–12 weeks+ of application. Normal detraining 1–2 weeks after non-attendance.	All populations.	Nil
Yoga	Body conditioning, health-related fitness benefits, postural and spinal improvements and improvement in flexibility/ mobility. Benefiting all ages, especially sedentary and elderly individuals. Improved flexibility, postural control and strength measures after regular attendance (>8 weeks).	1–3 sessions per week. Even 1 session per week has demonstrated improvements. >60min classes × 2/week or 90min classes × 1/ week. Research demonstrating benefits from 8–12 weeks+ of application and adaptation.	All populations.	Nil

Stretching Modality or Presentation	Rational and Outcome	Prescription	Population	DOMS Benefit?
Acute Injuries	Improve post-injury ROM. ROM will be reduced as a result of swelling and bleeding around the injury site, so this normal functional range needs to be regained. Improve post-injury ROM and return back to pre-injury functional range. The subsequent realignment of collagen and graduated load and stress applied from stretching and load to the injury site will lead to a more positive outcome during rehabilitation, as long as appropriate strength progressions and load are placed upon the injury site. Consideration of pain-free movement, range and progressive tolerance to load can help success.	Static-active stretching, progressing to static-passive stretching relative to injury site, progressing to more functional movements, isometric, concentric and eccentric training, and sport-specific ROM demands. Sets, repetitions and frequency: as clinically appropriate; pain-free ROM. Acute phase 4 x 30 seconds daily to regain pre-injury ROM. Long-term benefit: return to pre-injury ROM and full pre-injury function with reduced risk of re-injury.	All relevant (acute injury) populations.	Nil

Stretching Modality or Presentation	Rational and Outcome	Prescription	Population	DOMS Benefit?
Post-Operative and Chronic Pain	Improve post-surgical ROM. ROM will be reduced as a result of swelling, bleeding and scar-tissue formation around the post-operative site, so a normal functional ROM needs to be regained. Pain management of chronic pain, with improved ROM and stretch tolerance, and a return to normal function in combination with appropriate manual therapy. Improved ROM and return to pre-operative functional ROM. Reduction in chronic pain.	Static-active stretching, progressing to static-passive stretching relative to post-operative site and/ or associated area/tissues in chronic pain. Global posture re-education has been shown to have a similar outcome to traditional conventional stretching, but the practitioner needs to have additional training to apply this form of stretching. Sets: as clinically appropriate; 2–3 sets within a pain-free ROM. 30sec static holds. Frequency: as clinically appropriate; certainly regular/daily exposure. Long-term benefit: return to pre-surgical ROM and full function; reduction in chronic pain; improved self-esteem.	All relevant (post-operative-chronic pain) populations.	Nil
Adverse Neural Tension	Reduce neural adhesions and encourage gliding through the neural system. To be completed by those with a positive slump test. Positive slump test progresses to a negative slump test and full pain-free ROM with zero neural tension.	Mode: slump position as completed in a neural tension assessment. 3 sets of 8–12reps left and right side in a slump position. Alternate sides. Completed before exercise and also possibly on rest days. Long-term benefit: mobilize adhesions within the neural tissue to help reduce neural symptoms and improve reduced ROM. Important post-hamstring strain to reduce incidence of re-injury.	All populations when clinically indicated. (Positive slump)	Nil

Stretching Modality or Presentation	Rational and Outcome	Prescription	Population	DOMS Benefit?
Growth/ Age-related Injury	Maintain tissue extensibility as growth occurs to prevent growth injury in youth populations. Reduced incidence of growth-related injury.	Stitching modality selected from: PNF Myofascial release Vibration training All prescribed as previously directed to increase ROM with no compromise on performance and known muscle length gains. Group Pilates and/or Yoga sessions may ale be beneficial in this cohort	Youths experiencing a growth spurt or involved in regular exercise/sport which may predispose growth-related injury. May be prescribed prophylactic-ally for prevention.	Nil

The application of each stretching modality is hopefully explained thoroughly and clarified within this book to allow for appropriate application and prescription in whichever setting you may be in and whichever population you may be working with. I hope that you all have a positive outcome from the modality of choice!

USEFUL WEBSITES

www.brianmac.co.uk (BrianMac Sports Coach)

www.functionalmovement.com (Functional Movement Systems)

www.cooperinstitute.org/youth/fitnessgram (FitnessGram information)

www.hypermobility.org (Beighton Score and hypermobility information)

f-marc.com/11plus/11plus (the FIFA 11+ is a complete warm-up programme to reduce injuries among male and female football players aged fourteen years and older)

www.british-gymnastics.org (national governing body for Gymnastics in Britain)

www.physioroom.com (provides a wide range of news, guides and advice related to injuries in sport, plus an extensive range of injury products, physiotherapy supplies and rehabilitation equipment)

www.uksca.org.uk (United Kingdom Strength and Conditioning Association for strength and conditioning research and information)

www.nsca.com (the National Strength and Conditioning Association is the world-leading membership organization for thousands of elite strength coaches, personal trainers and dedicated researchers and educators)

www.csp.org.uk (the Chartered Society of Physiotherapy is the professional, educational and trade union body for the UK's 54,000 chartered physiotherapists, physiotherapy students and support workers)

www.nata.org (The professional membership association for certified athletic bodies)

GLOSSARY

Active stretching Active stretching is performed without any aid or assistance from an external force. You stretch a muscle by actively contracting the muscle in opposition to the one you're stretching.

Aerobic capacity (VO2 Max) The ability to take in and use oxygen during exercise. The amount of oxygen a person can utilize during maximal exercise is determined by cardiovascular conditioning and the efficiency of oxygen extraction around the body during exercise.

Agonists and antagonist muscles A contracting muscle whose contraction is opposed by another muscle (an antagonist). The antagonist muscle counteracts the action of the agonist. For example, biceps and triceps work together as agonist and antagonist muscles.

Apophysis Any natural protrusion forming part of a bone, such as a tubercle or tuberosity.

Apophysitis Inflammation of an apophysis. Apophysitis occurs due to excessive traction or stress most frequently affecting the calcaneus (Sever's disease), the knee (Osgood-Schlatter), the shoulder (Little Leaguer shoulder) or elbow (Little Leaguer elbow).

Calisthenics Systematic rhythmic bodily exercises, similar to gymnastic movements, performed usually without apparatus.

Cardiac muscle (heart muscle) An involuntary, striated muscle that is found in the walls of the heart, specifically the myocardium.

Concentric muscle action Muscle contraction resulting in shortening of a muscle.

Contractures A permanent shortening of non-bony tissues like muscles, tendons, or scar tissue producing deformity or distortion.

Delayed Onset Muscle Soreness (DOMS) The muscular pain and soreness felt several hours to days after unaccustomed or strenuous exercise. The soreness is typically more intense around 24 to 72 hours after exercise.

Eccentric muscle action Elongation of a muscle where the muscle acts to decelerate the joint at the end of a movement or control the repositioning of a load.

Endomysium The connective tissue surrounding the individual muscular fibres within the smallest bundles. It is the deepest and smallest component of connective tissue.

Epimysium The external connective-tissue sheath of a muscle that protects muscles from friction against other muscles and bones.

Extensible Capable of being stretched and ability to accommodate change.

Fascia A sheet of connective tissue surrounding muscles, bones, organs, nerves, blood vessels and other structures within the body.

Flexibile Capable of being turned, bowed, or twisted without breaking.

Golgi tendon organs A proprioceptive sensory nerve ending embedded among the fibres of a tendon, often near the musculotendinous junction; it is activated by any increase of tension in

149

the tendon, caused either by active contraction or passive stretch of the corresponding muscle.

Grade I Muscle Strain The muscle or tendon is overstretched. Small tears to muscle fibres may or may not occur. There may be mild pain with or without swelling.

Grade II Muscle Strain The muscle or its tendon is overstretched with more of the fibres torn but not completely ruptured. Symptoms may include pain with swelling. The injury site is tender. Bruising may occur if small blood vessels at the site of injury are damaged as well. Movement may be difficult because of pain and swelling.

Grade III Muscle Strain In many cases, many or indeed all of the muscle fibres are torn or ruptured. Pain, swelling, tenderness and bruising are usually present. Movement is usually very difficult.

Isometric muscle action Muscle contraction without a shortening or change in length between its origin and insertion.

Length-tension relationship The relationship between a muscle's length and the isometric tension (force) which it generates when fully activated.

Mobile Capable of moving freely or being moved about readily.

Myofibril (also known as a muscle fibril) One of the longitudinal parallel contractile elements of a muscle cell that are composed of myosin and actin. Muscles contract by sliding the thick (myosin) and thin (actin) filaments along each other.

Passive stretching To use some sort of outside assistance to help you achieve a stretch. This assistance could be your body weight, a strap, a step, gravity, a partner or a stretching device.

Perimysium The connective-tissue sheath that surrounds a muscle and forms sheaths for the bundles of muscle fibres.

Pilates An exercise regime typically performed with the use of specialized apparatus and designed to improve the overall condition of the body.

Proprioceptors Allow your body to know where it is in space. Sensory receptors in muscles, joint capsules and surrounding tissues will signal information to the central nervous system about the position and movement of body parts. An example would be the angle at a joint or the length of a muscle.

Range of movement (ROM) The measurement of movement around a specific joint or body part.

Sarcomere The basic unit of striated muscle tissue. Sarcomeres are composed of long, fibrous proteins as filaments that slide past each other when a muscle contracts or relaxes.

Skeletal muscle A form of striated muscle that is usually attached to the skeleton and is usually under voluntary control.

Sliding filament theory The actin (thin) filaments of muscle fibres slide past the myosin (thick) filaments during muscle contraction, while the two groups of filaments remain at relatively constant length. These muscle proteins slide past each other to generate movement.

Smooth muscle An involuntary non-striated muscle. Smooth muscle is found within the walls of blood vessels, lymphatic vessels, the urinary bladder, uterus, male and female reproductive tracts and the gastrointestinal tract.

Stretch To extend in length or to become extended in either length or breadth.

Stretch Shortening Cycle (SSC) The 'pre-stretch' action that is commonly observed during typical human movements such as jumping and bounding. An eccentric muscle contraction followed immediately by a concentric contraction of the same muscle group is an example of this pre-stretch which allows the athlete to produce more force and move quicker and can be trained with plyometric (jumping) based activities.

Trigger Point A localized and usually tender or painful area of the body, usually a muscle, that when stimulated causes generalised musculoskeletal pain.

Yoga A system of physical postures, breathing techniques, and meditation to promote bodily or mental control and well-being.

ENDNOTES

1 *Oxford English Dictionary* (Oxford: Oxford University Press, 2015).

2 Fascia Research Congress (2009), 'Terminology Used in Fascia Research'; available from http://www.fasciacongress.org/2009/glossary.htm [accessed 5 April 2012].

3 BrianMac Sports Coach, http://www.brianmac.co.uk.

4 Weppler, C.H. and Magnusson, S.P., 'Increasing Muscle Extensibility: A Matter of Increasing Length or Modifying Sensation?', *Physical Therapy*, March 2010, 90 (3), pp. 438–49.

5 Özkaya, N. and Nordin, M., *Fundamentals of Biomechanics: Equilibrium, Motion and Deformation*, 2nd Edition (New York: Springer; 1999).

6 Magnusson, S.P., 'Passive Properties of Human Skeletal Muscle During Stretch Manoeuvres: A Review', *Scandinavian Journal of Medicine and Science in Sports*, 1998, 8, pp. 65–77.

7 Weppler and Magnusson, *op. cit.* in note 4.

8 Chan, S.P., Hong Y. and Robinson P.D., 'Flexibility and Passive Resistance of the Hamstrings of Young Adults Using Two Different Static Stretching Protocols', *Scandinavian Journal of Medicine and Science in Sports*, 2001, 11, pp. 81–6.

9 Draper, D.O., Castro, J.L., Feland, B. *et al.*, 'Shortwave Diathermy and Prolonged Stretching Increase Hamstring Flexibility More than Prolonged Stretching Alone', *Journal of Orthopaedic and Sports Physical Therapy*, 2004, 34, pp. 13–20.

10 Weppler and Magnusson, *op. cit.* in note 4.

11 Tabary, J.C., Tabary, C., Tardieu, C., Tardieu, G. and Goldspink, G., 'Physiological and Structural Changes in the Cat's Soleus Muscle Due to Immobilization at Different Lengths by Plaster Casts', *The Journal of Physiology*, July 1972, 224 (1), pp. 231–44.

12 Weppler and Magnusson, *op. cit.* in note 4.

13 Halbertsma, J.P.K. and Göeken, L.N.H, 'Stretching Exercises: Effect on Passive Extensibility and Stiffness in Short Hamstrings of Healthy Subjects', *Archives of Physical Medicine and Rehabilitation*, September 1994, 75 (9), pp. 976–81.

14 Kraus, K., Schütz, E., Taylor, W.R. and Doyscher, R., 'Efficacy of the Functional Movement Screen: A Review', *The Journal of Strength and Conditioning Research*, December 2014, 28 (12), pp. 3,571–84.

15 Wells, K.F. and Dillon, E.K., 'The Sit and Reach. A Test of Back and Leg Flexibility', *Research Quarterly*, 1952, 23, pp. 115–8.

16 Holt *et al.*, 'Modifications to the Standard Sit-and-Reach Flexibility Protocol', *Journal of Athletic Training*, 1999, 34 (1), pp. 43–7.

17 Burke, D.D., *A Comparison of Manual and Machine Assisted PNF Flexibility Techniques* [master's thesis] (Halifax, Nova Scotia: Dalhousie University, 1994).

18 Gyoung-Mo Kim and Sung-Min Ha, 'Reliability of the Modified Thomas Test Using a Lumbo-pelvic Stabilization', *Journal of Physical Therapy Science*, February 2015, 27 (2), pp. 447–9.

19 Peeler, J.D. and Anderson, J.E., 'Reliability Limits of the Modified Thomas Test for Assessing Rectus Femoris Muscle Flexibility About the Knee Joint', *Journal of Athletic Training*, September–October 2008, 43 (5), pp. 470–6, doi: 10.4085/1062-6050-43.5.470.

20 Pope, R., Herbert, R. and Kirwan, J., 'Effect of Ankle Dorsiflexion Range and Pre-exercise Calf Muscle Stretching on Injury Risk in Army Recruits', *Australian Physiotherapy*, 44 (3), pp. 165–72.

21 Gabbe, B.J., Finch, C.F., Wajswelner, H. and Bennell, K.L., 'Predictors of Lower Extremity Injuries at the Community Level of Australian Football', *Clinical Journal of Sport Medicine*, 2004, 14 (2), pp. 56–63.

22 Bennell, K.L., Talbot, R., Wajswelner, H., Techovanich, W. and Kelly, D., 'Intra-rater and Inter-tester Reliability of a Weightbearing Lunge Measure of Ankle Dorsiflexion', *Australian Physiotherapy*, 1998, 24 (2), pp. 211–17.

23 De Araújo, C.G., 'Flexibility Assessment: Normative Values for Flexitest from 5 to 91 Years of Age', *Arquivos Brasileiros de Cardiologia*, April 2008, 90 (4), pp. 257–63.

24 De Araújo, C.G., *Flexitest: An Innovative Flexibility Assessment Method* 1st Edition (Champaign, IL: Human Kinetics, 2003).

25 Herbert R.D., de Noronha, M. and Kamper, S.J., 'Stretching to Prevent or Reduce Muscle Soreness after Exercise', *Cochrane Database of Systematic Reviews,* 2011, 7, Art. No.: CD004577. DOI: 10.1002/14651858.CD004577.pub3.

26 Siatras, T.A., Mittas, V.P., Mameletzi, D.N. and Vamvakoudis, E.A., 'The Duration of the Inhibitory Effects with Static Stretching on Quadriceps Peak Torque Production', *The Journal of Strength and Conditioning Research*, January 2008, 22 (1), pp. 40–6.

27 Ogura, Y., Miyahara, Y., Naito, H., Katamoto, S. and Aoki, J., 'Duration of Static Stretching Influences Muscle Force Production in Hamstring Muscles', *The Journal of Strength and Conditioning Research*, August 2007, 21 (8), pp. 788–92.

28 Beckett, J.R., Schneiker, K.T., Wallman, K.E., Dawson, B.T. and Guelfi, K.J., 'Effects of Static Stretching on Repeated Sprint and Change of Direction Performance', *Medicine and Science in Sports and Exercise*, February 2009, 41 (2), pp. 444–50.

29 Kokkonen, J., Nelson, A.G., Eldredge, C. and Winchester, J.B., 'Chronic Static Stretching Improves Exercise Performance', *Medicine and Science in Sports and Exercise*, October 2007, 39 (10), pp. 1,825–31.

30 Behm, D.G., Bambury, A., Cahill, F. and Power, K., 'Effect of Acute Static Stretching on Force, Balance, Reaction Time, and Movement Time', *Medicine and Science in Sports and Exercise*. August 2004, 36 (8), pp. 1,397–402.

31 Allison, S.J., Bailey, D.M. and Folland, J.P., 'Prolonged Static Stretching Does Not Influence Running Economy Despite Changes in Neuromuscular Function', *Journal of Sports Sciences*, December 2008, 26 (14), pp. 1,489–95.

32 Lowery, R.P., Joy, J.M., Brown, L.E., Oliveira de Souza, E., Wistocki, D.R., Davis, G.S., Naimo, M.A., Zito, G.A. and Wilson, J.M., 'Effects of Static Stretching on 1-mile Uphill Run Performance',

The Journal of Strength and Conditioning Research, January 2014, 28 (1), pp. 161–7.

33 Wilson, J.M., Hornbuckle, L.M., Kim, J.S., Ugrinowitsch, C., Lee, S.R., Zourdos, M.C., Sommer, B. and Panton, L.B., 'Effects of Static Stretching on Energy Cost and Running Endurance Performance', *The Journal of Strength and Conditioning Research*, September 2010, 24 (9), pp. 2,274–9.

34 Yamaguchi, T., Ishii, K., Yamanaka, M. and Yasuda, K., 'Acute Effect of Static Stretching on Power Output During Concentric Dynamic Constant External Resistance Leg Extension', *The Journal of Strength and Conditioning Research*, November 2006, 20 (4), pp. 804–10.

35 DePino, G.M., Webright, W.G. and Arnold, B.L., 'Duration of Maintained Hamstring Flexibility after Cessation of an Acute Static Stretching Protocol', *Journal of Athletic Training*, January–March 2000, 35 (1), pp. 56–9.

36 Kokkonen et al., *op. cit.* in note 29.

37 Herbert, R.D., de Noronha, M. and Kamper, S.J., 'Stretching to Prevent or Reduce Muscle Soreness after Exercise', *Cochrane Database of Systematic Reviews*, July 2011, 6 (7), CD004577.

38 Cheung, K., Hume, P. and Maxwell, L., 'Delayed Onset Muscle Soreness: Treatment Strategies and Performance Factors', *Sports Medicine*, 2003, 33 (2), pp. 145–64.

39 Ogura et al., *op. cit.* in note 27.

40 Beckett et al., *op. cit.* in note 28.

41 Kokkonen et al., *op. cit.* in note 29.

42 Wilson et al., *op. cit.* in note 33.

43 Siatras et al., *op. cit.* in note 26.

44 Behm et al., *op. cit.* in note 30.

45 Lowery et al., *op. cit.* in note 32.

46 DePino et al., *op. cit.* in note 35.

47 Herbert et al., *op. cit.* in note 37.

48 Samson, M., Button, D.C., Chaouachi, A. and Behm. D.G., 'Effects of Dynamic and Static Stretching within General and Activity Specific Warm-up Protocols', *The Journal of Sports Science and Medicine*, June 2012, 11 (2), pp. 279–85.

49 Mascarin, N.C., Vancini, R.L., Lira, C.A. and Andrade, M.S., 'Stretch-induced Reductions in Throwing Performance are Attenuated by Warm-up before Exercise', *The Journal of Strength and Conditioning Research*, November 2014, 28 (11).

50 Zourdos, M.C., Wilson, J.M., Sommer, B.A., Lee, S.R., Park, Y.M., Henning, P.C., Panton, L.B. and Kim, J.S., 'Effects of Dynamic Stretching on Energy Cost and Running Endurance Performance in Trained Male Runners', *The Journal of Strength and Conditioning Research*, February 2012, 26 (2), pp. 335–41.

51 Wilson et al., *op. cit.* in note 33.

52 Wilson et al., *op. cit.* in note 33.

53 Behm, D.G. and Chaouachi, A., 'A Review of the Acute Effects of Static and Dynamic Stretching on Performance', *European Journal of Applied Physiology*, November 2011, 111 (11), pp. 2,633–51.

54 Witvrouw, E., Mahieu, N., Danneels, L. and McNair, P., 'Stretching and Injury Prevention: An Obscure Relationship, *Sports Medicine*, 2004, 34 (7), pp. 443–9.

55 Perrier, E.T., Pavol, M.J. and Hoffman, M.A., 'The Acute Effects of a Warm-up Including Static or Dynamic Stretching on Countermovement Jump Height, Reaction Time, and Flexibility', *The Journal of Strength and Conditioning Research*, July 2011, 25 (7), pp. 1,925–31.

56 Needham, R.A., Morse, C.I. and Degens, H.J., 'The Acute Effect of Different Warm-up Protocols on Anaerobic Performance in Elite Youth Soccer Players', *The Journal of Strength and Conditioning Research*, December 2009, 23 (9), pp. 2,614–20.

57 Yamaguchi, T. and Ishii, K., 'Effects of

Static Stretching for 30 Seconds and Dynamic Stretching on Leg Extension Power', *The Journal of Strength and Conditioning Research*, August 2005, 19 (3), pp. 677–83.

58 Needham *et al.*, *op. cit.* in note 56.

59 Bacurau, R.F., Monteiro, G.A., Ugrinowitsch, C., Tricoli, V., Cabral, L.F. and Aoki, M.S., 'Acute Effect of a Ballistic and a Static Stretching Exercise Bout on Flexibility and Maximal Strength, *The Journal of Strength and Conditioning Research*, January 2009, 23 (1), pp. 304–8.

60 Jaggers, J.R., Swank, A.M., Frost, K.L. and Lee, C.D., 'The Acute Effects of Dynamic and Ballistic Stretching on Vertical Jump Height, Force, and Power', *The Journal of Strength and Conditioning Research*, November 2008, 22 (6), pp. 1,844–9.

61 Woolstenhulme, M.T., Griffiths, C.M., Woolstenhulme, E.M. and Parcell, A.C., 'Ballistic Stretching Increases Flexibility and Acute Vertical Jump Height When Combined with Basketball Activity', *The Journal of Strength and Conditioning Research*, November 2006, 20 (4).

62 Sharman, M.J., Cresswell, A.G. and Riek, S., 'Proprioceptive Neuromuscular Facilitation Stretching: Mechanisms and Clinical Implications', *Sports Medicine*, 2006, 36 (11), pp. 929–39.

63 Chalmers, G., 'Re-examination of the Possible Role of Golgi Tendon Organ and Muscle Spindle Reflexes in Proprioceptive Neuromuscular Facilitation Muscle Stretching', *Sports Biomechanics*, January 2004, 3 (1), pp. 159–83.

64 Higgs, F. and Winter, S.L., 'The Effect of a Four-week Proprioceptive Neuromuscular Facilitation Stretching Program on Isokinetic Torque Production', *The Journal of Strength and Conditioning Research*, August 2009, 23 (5), pp. 1,442–7.

65 O'Hora, J., Cartwright, A., Wade, C.D.,

Hough A.D. and Shum, G.L, 'Efficacy of Static Stretching and Proprioceptive Neuromuscular Facilitation Stretch on Hamstring Length after a Single Session', *The Journal of Strength and Conditioning Research*, June 2011, 25 (6), pp. 1,586–91.

66 Kwak, D.H. and Ryu, Y.U., 'Applying Proprioceptive Neuromuscular Facilitation Stretching: Optimal Contraction Intensity to Attain the Maximum Increase in Range of Motion in Young Males', *Journal of Physical Therapy Science*, July 2015, 27 (7), pp. 2,129–32, doi: 10.1589/jpts.27.2129; Epub 22 July 2015.

67 Grieve, R., Goodwin, F., Alfaki, M., Bourton, A.J., Jeffries, C. and Scott, H., 'The Immediate Effect of Bilateral Selfmyofascial Release on the Plantar Surface of the Feet on Hamstring and Lumbar Spine Flexibility: A Pilot Randomised Controlled Trial', *Journal of Bodywork and Movement Therapies*, July 2015, 19 (3), pp. 544–52.

68 Couture, G., Karlik, D., Glass, S.C. and Hatzel, B.M., 'The Effect of Foam Rolling Duration on Hamstring Range of Motion', *The Open Orthopaedics Journal*, 2 October 2015, 9, pp. 450–5.

69 Junker, D.H. and Stöggl, T.L., 'The Foam Roll as a Tool to Improve Hamstring Flexibility', *The Journal of Strength and Conditioning Research*, December 2015, 29 (12), pp. 3,480–5.

70 MacDonald, G.Z., Penney, M.D., Mullaley, M.E., Cuconato, A.L., Drake, C.D., Behm, D.G. and Button, D.C., 'An Acute Bout of Self-myofascial Release Increases Range of Motion without a Subsequent Decrease in Muscle Activation or Force', *The Journal of Strength and Conditioning Research*, March 2013, 27 (3), pp. 812–21.

71 Healey, K.C., Hatfield, D.L., Blanpied, P., Dorfman, L.R. and Riebe, D., 'The Effects of Myofascial Release with Foam Rolling

on Performance', *The Journal of Strength and Conditioning Research*, January 2014, 28 (1), pp. 61–8.

72 Pearcey, G.E., Bradbury-Squires, D.J., Kawamoto, J.E., Drinkwater, E.J., Behm, D.G. and Button, D.C., 'Foam Rolling for Delayed-onset Muscle Soreness and Recovery of Dynamic Performance Measures', *Journal of Athletic Training*, January 2015, 50 (1), pp. 5–13, doi: 10.4085/1062-6050-50.1.01; Epub 21 November 2014.

73 Schroeder, A.N. and Best, T.M., 'Is Self-myofascial Release an Effective Pre-exercise and Recovery Strategy?', *A Literature Review. Current Sports Medicine Reports*, May–June 2015, 14 (3), pp. 200–8.

74 Dallas, G., Paradisis, G., Kirialanis, P., Mellos, V., Argitaki, P. and Smirniotou, A., 'The Acute Effects of Different Training Loads of Whole Body Vibration on Flexibility and Explosive Strength of Lower Limbs', *Biology of Sport*, September 2015, 32 (3), pp. 235–41.

75 Sands, W.A., McNeal, J.R., Stone, M.H., Haff, G.G. and Kinser, A.M., 'Effect of Vibration on Forward Split Flexibility and Pain Perception in Young Male Gymnasts', *International Journal of Sports Physiology and Performance*, December 2008, 3 (4), pp. 469–81.

76 Fagnani, F., Giombini, A., Di Cesare, A., Pigozzi, F. and Di Salvo, V., 'The Effects of a Whole-body Vibration Program on Muscle Performance and Flexibility in Female Athletes', *American Journal of Physical Medicine & Rehabilitation*, December 2006, 85 (12), pp. 956–62.

77 Phrompaet, S., Paungmali, A., Pirunsan, U. and Sitilertpisan, P., 'Effects of Pilates Training on Lumbo-pelvic Stability and Flexibility', *Asian Journal of Sports Medicine*, March 2011, 2 (1), pp. 16–22.

78 Kibar, S., Yardimci, F.O., Evcik, D., Ay, S., Alhan, A., Manço, M. and Ergin, E.S., 'Is Pilates Exercise Program Effective on Balance, Flexibility and Muscle Endurance? Randomized, Controlled Study', *The Journal of Sports Medicine and Physical Fitness*, 16 October 2015.

79 Vaquero-Cristóbal, R., López-Miñarro, P.A., Alacid Cárceles, F. and Esparza-Ros, F., 'The Effects of the Pilates Method on Hamstring Extensibility, Pelvic Tilt and Trunk Flexion', *Nutrición Hospitalaria*, November 2015, 32 (5), pp. 1,967–86.

80 Kloubec, J.A., 'Pilates for Improvement of Muscle Endurance, Flexibility, Balance, and Posture', *The Journal of Strength and Conditioning Research*, March 2010, 24 (3), pp. 661–7.

81 Farinatti, P.T., Rubini, E.C., Silva, E.B. and Vanfraechem, J.H., 'Flexibility of the Elderly after One-year Practice of Yoga and Calisthenics', *International Journal of Yoga Therapy*, September 2014, pp. 71–7.

82 De Araújo, *op. cit.* in note 23.

83 Tran, M.D., Holly, R.G., Lashbrook, J. and Amsterdam, E.A., 'Effects of Hatha Yoga Practice on the Health-Related Aspects of Physical Fitness', *Preventative Cardiology*, Autumn 2001, 4 (4), pp. 165–70.

84 Gothe, N.P. and McAuley, E., 'Yoga is as Good as Stretching-Strengthening Exercises in Improving Functional Fitness Outcomes: Results from a Randomized Controlled Trial', *The Journals of Gerontology. Series A, Biological Sciences and Medical Sciences,* 22 August 2015, pii: glv127.

85 Sherman, K.J., Cherkin, D.C., Wellman, R.D., Cook, A.J., Hawkes, R.J., Delaney, K. and Deyo, R.A., 'A Randomized Trial Comparing Yoga, Stretching, and a Self-care Book for Chronic Low Back Pain', *Archives of Internal Medicine*, December 2011, 171 (22), pp. 2,019–26.

86 Grabara, M. and Szopa, J., 'Effects of Hatha Yoga Exercises on Spine Flexibility in Women over 50 Years Old', *Journal of*

Physical Therapy Science, February 2015, 27 (2), pp. 361–5.

87 Adirim, T.A. and Cheng, T.L., 'Overview of Injuries in the Young Athlete', *Sports Medicine*, 2003, 33 (1), pp. 75–81.

88 Nakase, J., Goshima, K., Numata, H., Oshima, T. and Takata, Y., 'Tsuchiya H Precise Risk Factors for Osgood-Schlatter Disease', *Archives of Orthopaedic and Trauma Surgery*, September 2015, 135 (9), pp. 1,277–81.

89 Ehrenborg, G. and Lagergren, C., 'Roentgenologic Changes in the Osgood-Schlatter Lesion', *Acta Chirurgica Scandinavica*, 1961, 121, pp. 315–27.

90 Antich, T.J. and Brewster, C.E., 'Osgood-Schlatter Disease: Review of Literature and Physical Therapy Management', *Journal of Orthopaedic & Sports Physical Therapy*, 1985, 7 (1), pp. 5–10.

91 Elengard, T., Karlsson, J. and Silbernagel, K.G., 'Aspects of Treatment for Posterior Heel Pain in Young Athletes', *Open Access Journal of Sports Medicine*, 2010, 1, pp. 223–32.

92 Park, D.Y., Rubenson, J., Carr, A., Mattson, J., Besier, T. and Chou, L.B., 'Influence of Stretching and Warm-up on Achilles Tendon Material Properties', *Foot & Ankle International*, April 2011, 32 (4), pp. 407–13.

93 Park, D.Y. and Chou, L., 'Stretching for Prevention of Achilles Tendon Injuries: A Review of the Literature', *Foot & Ankle International*, December 2006, 27 (12), pp. 1,086–95.

94 Witvrouw, E., Mahieu, N., Roosen, P. and McNair, P., 'The Role of Stretching in Tendon Injuries', *British Journal of Sports Medicine*, April 2007, 41 (4), pp. 224–6.

95 Mahieu, N.N., McNair, P., De Muynck, M., Stevens, V., Blanckaert, I., Smits, N. and Witvrouw, E., 'Effect of Static and Ballistic Stretching on the Muscle-tendon Tissue Properties', *Medicine and Science in Sports and Exercise*, 2007, 39 (3), pp. 494–501.

96 Kamonseki, D.H., Gonçalves, G.A., Yi, L.C. and Júnior, I.L., 'Effect of Stretching with and without Muscle Strengthening Exercises for the Foot and Hip in Patients with Plantar Fasciitis: A Randomized Controlled Single-blind Clinical Trial', *Manual Therapy*, 30 October 2015, pii: S1356-689X(15)00196-4, doi: 10.1016/j.math.2015.10.006. [Epub ahead of print]

97 Celik, D., Kus., G. and Sırma, S.Ö., 'Joint Mobilization and Stretching Exercise vs Steroid Injection in the Treatment of Plantar Fasciitis: A Randomized Controlled Study', *Foot & Ankle International*, 23 September 2015, pii: 1071100715607619. [Epub ahead of print.]

98 Behm, D.G., Blazevich, A.J., Kay, A.D. and McHugh, M., 'Acute Effects of Muscle Stretching on Physical Performance, Range of Motion, and Injury Incidence in Healthy Active Individuals: A Systematic Review', *Applied Physiology, Nutrition and Metabolism*, 8 December 2015, pp. 1–11. [Epub ahead of print.]

99 Barengo, N.C., Meneses-Echávez, J.F., Ramírez-Vélez, R., Cohen, D.D., Tovar, G. and Bautista, J.E., 'The Impact of the FIFA 11+ Training Program on Injury Prevention in Football Players: A Systematic Review', *International Journal of Environmental Research and Public Health*, 19 November 2014, 11 (11), pp. 11,986–2,000.

100 Lauersen, J.B., Bertelsen, D.M. and Andersen, L.B., 'The Effectiveness of Exercise Interventions to Prevent Sports Injuries: A Systematic Review and Meta-analysis of Randomised Controlled Trials', *British Journal of Sports Medicine*, June 2014, 48 (11), pp. 871–7.

101 Järvinen, T.A.H., Markku, J. and Kalimo, H., 'Regeneration of Injured Skeletal Muscle after the Injury', *Muscles, Ligaments and Tendons Journal*, October–December 2013, 3 (4), pp. 337–45.

102 Sherry, M.A. and Best, T.M., 'A Comparison of 2 Rehabilitation Programs in the Treatment of Acute Hamstring Strains', *Journal of Orthopaedic & Sports Physical Therapy*, March 2004, 34 (3), pp. 116–25.

103 Silder, A., Sherry, M.A., Sanfilippo, J., Tuite, M.J., Hetzel, S.J. and Heiderscheit B.C., 'Clinical and Morphological Changes Following 2 Rehabilitation Programs in the Treatment of Acute Hamstring Strain Injuries: A Randomized Clinical Trial', *Journal of Orthopaedic & Sports Physical Therapy*, May 2013, 43 (5), pp. 284–99.

104 Malliaropoulos, N., Papalexandris, S., Papalada, A. and Papacostas, E., 'The Role of Stretching in Rehabilitation of Hamstring Injuries: 80 Athletes Follow-up', *Medicine & Science in Sports Exercise*, May 2004, 36 (5), pp. 756–9.

105 Valle, X., Tol, J.L., Hamilton, B., Rodas, G., Malliaras, P., Malliaropoulos, N., Rizo, V., Moreno, M. and Jardi, J., 'Hamstring Muscle Injuries, a Rehabilitation Protocol Purpose', *Asian Journal of Sports Medicine*, December 2015, 6 (4), pp. e25,411.

106 Nsitem, V. 'Diagnosis and Rehabilitation of Gastrocnemius Muscle Tear: A Case Report', *Journal of the Canadian Chiropractic Association*, December 2013, 57 (4), pp. 327–33.

107 O'Sullivan, K., McAuliffe, S. and Deburca, N., 'The Effects of Eccentric Training on Lower Limb Flexibility: A Systematic Review', *British Journal of Sports Medicine*, September 2012, 46 (12), pp. 838–45.

108 Ayhan, H., Tastan, S., Iyigün, E., Oztürk, E., Yildiz, R. and Görgülü, S., 'The Effectiveness of Neck Stretching Exercises Following Total Thyroidectomy on Reducing Neck Pain and Disability: A Randomized Controlled Trial', *Worldviews on Evidence-Based Nursing*, 15 January 2016, doi: 10.1111/wvn.12136. [Epub ahead of print.]

109 Lee, T.S., Kilbreath, S.L., Refshauge, K.M., Pendlebury, S.C., Beith, J.M. and Lee, M.J., 'Pectoral Stretching Program for Women Undergoing Radiotherapy for Breast Cancer', *Breast Cancer Research and Treatment*, May 2007, 102 (3), pp. 313–21.

110 Castagnoli, C., Cecchi, F., Del Canto, A., Paperini, A., Boni, R., Pasquini, G., Vannetti, F. and Macch, C., 'Effects in Short and Long Term of Global Postural Reeducation (GPR) on Chronic Low Back Pain: A Controlled Study with One-Year Follow-Up', *Scientific World Journal*, 2015.

111 Cunha, A.C.V., Burke, T.N., França, F.J.R. and Marques, A.P., 'Effect of Global Posture Re-education and of Static Stretching on Pain, Range of Motion, and Quality of Life in Women with Chronic Neck Pain: A Randomized Clinical Trial', *Clinics*, December 2008, 63 (6), pp. 763–70.

112 Katalinic, O.M., Harvey, L.A. and Herbert, R.D., 'Effectiveness of Stretch for the Treatment and Prevention of Contractures in People with Neurological Conditions: A Systematic Review', *Physical Therapy*, January 2011, 91 (1), pp. 11–24.

113 McHugh, M.P., Johnson, C.D. and Morrison, R.H., 'The Role of Neural Tension in Hamstring Flexibility', *Scandinavian Journal of Medicine & Science in Sports*, April 2012, 22 (2), pp. 164–9.

114 Kornberg, C. and Lew, P., 'The Effect of Stretching Neural Structures on Grade One Hamstring Injuries', *Journal of Orthopaedic & Sports Physical Therapy*, October 1989, 10 (12), pp. 481–7.

115 Turl, S.E. and George, KP., 'Adverse Neural Tension: A Factor in Repetitive Hamstring Strain?', *Journal of Orthopaedic & Sports Physical Therapy*, January 1998, 27 (1), pp. 16–21.

116 Miller, K.C., Stone, M.S., Huxel, K.C. and Edwards, J.E., 'Exercise-associated Muscle Cramps: Causes, Treatment, and Prevention', *Sports Health*, July 2010, 2 (4), pp. 279–83.

117 Miller, K.C. and Burne, J.A., 'Golgi Tendon Organ Reflex Inhibition Following Manually Applied Acute Static Stretching', *Journal of Sports Sciences*, 2014, 32 (15), pp. 1,491–7, doi: 10.1080/02640414.2014.899708, Epub 9 April 2014.

118 DePino et al., op. cit. in note 35.

119 Bentley, S., 'Exercise-induced Muscle Cramp. Proposed Mechanisms and Management', *Sports Medicine*, June 1996, 21 (6), pp. 409–20.

120 Behringer, M., Moser, M., McCourt, M., Montag, J. and Mester, J., 'A Promising Approach to Effectively Reduce Cramp Susceptibility in Human Muscles: A Randomized, Controlled Clinical Trial', *PLoS One*, 11 April 2014, 9(4), e94910, doi: 10.1371/journal.pone.0094910, eCollection 2014.

121 Hallegraeff, J.M., van der Schans, C.P., de Ruiter, R. and de Greef, M.H.G., 'Stretching before Sleep Reduces the Frequency and Severity of Nocturnal Leg Cramps in Older Adults: A Randomised Trial, *Journal of Physiotherapy*, March 2012, 58 (1), pp. 17–22.

122 Coppin, R.J., Wicke, D.M. and Little, P.S., 'Managing Nocturnal Leg Cramps – Calf-Stretching Exercises and Cessation of Quinine Treatment: A Factorial Randomised Controlled Trial', *British Journal of General Practice*, March 2005, 55 (512), pp. 186–91.

123 Garrison, S.R., 'Prophylactic Stretching is Unlikely to Prevent Nocturnal Leg Cramps', *Journal of Physiotherapy*, September 2014, 60 (3), p. 174.

124 Suzuki, K., Miyamoto, M., Miyamoto, T. and Hirata, K., 'Sleep-Related Movement Disorders. *Nihon Rinsho*, June 2015, 73 (6), pp. 954–64 [Article in Japanese].

125 Franco, B.L., Signorelli, G.R., Trajano, G.S., Costa, P.B. and de Oliveira, C.G., 'Acute Effects of Three Different Stretching Protocols on the Wingate Test Performance', *Journal of Sports Science and Medicine*, March 2012, 11 (1), pp. 1–7.

INDEX